MILK
The Most Perfect Food

Extracts from Foreword by Pt. Madan Mohan Malviya

Vice-Chancellor,

Benares Hindu University

Dr. Godbole has rendered a service to the country by writing this book which he has done at my suggestion, as a means of propagating such knowledge...

He has dealt with the question both from the scientific and cultural point of view...

"Man does not live for bread alone. As a representative of the highest evolution of rational beings, he has to live for some ideals."...

The humanitarian in India looks upon the life of a cow that has yielded milk all her life-time with the same feeling with which he looks at the life of one of his own relatives and rightly too...

Both for the young and the adult, cow's milk is most valuable—the most perfect—of all human foods...

The illustrations give in the book are happily designed and show originality in the manner of their execution. They are meant to be at once educative, explanatory and convincing.

MILK
The Most Perfect Food

Prof. Dr. N.N. GODBOLE, M.A., B.Sc., Ph.D., (Berlin)
"Maharajah Sir Kirti Shah (Tehri Garhwal)"
Professor of Industrial Chemistry
Benares Hindu University

With a foreword by
PANDIT MADAN MOHAN MALVIYA
Vice-Chancellor,
Benares Hindu University

BIOTECH
2020
Biotech Books

ISBN 978-81-7622-179-5

Published by : **BIOTECH BOOKS**
4762-63/23, Ansari Road, Darya Ganj,
New Delhi - 110 002
Ph.: 011 43003222
e-mail: biotechbooks@yahoo.co.in

Laser Typesetting : **Classic Computer Services**
Delhi - 110 035

Printed at : **Chawla Offset Printers**
Delhi - 110 052

PRINTED IN INDIA

ACKNOWLEDGMENT.

It is a great pleasure to me to offer my thanks to many of my friends and students for the help they have rendered, in writing this book. Pt. S. D. Satawalekar, of Oundh State, has helped me with quotations from the Vedas. Dr. B. G. Ghanekar, of B. H. University, has supplied the references concerning milk, from the Ayurvedic books. Prof. B. P. Adarkar, M.A. (Oxon.) was of much help to me in revising the proofs. Messrs Sad Gopal, E. R. Bhide and N. T. Godbole have assisted me in writing out and typing the manuscript.

I have particularly to thank Mr. L. S. Wakankar, B. Sc. for the interesting and illustrative diagrams he has drawn in accordance with my suggestions and also for preparing the index.

Messrs. Wagle Process Studio (of J. Walter Thompson Company), Calcutta, have prepared the blocks and have done very good work.

The Manager, "Leader" Press deserves my special thanks for his ungrudging co-operation and help.

N. N. GODBOLE.

Benares Hindu University,

"Dipawali" (November) 1936.

FOREWORD.

National health is today receiving the foremost consideration of civilised governments in the world. It is everywhere recognised that it is the basic asset of a nation's welfare. All great countries are therefore adopting well-concerted and comprehensive measures for improving the health, physique, and power of endurance of their peoples.

No doubt heredity, climate and the social and political conditions under which a people live influence both physical and intellectual development. But when all these have been taken into account, it remains true that national health depends very largely upon the national diet. Consequently in the advanced countries of the world, the Science of Nutrition has received much attention during the last few decades, and particularly so since the time of the Great War. Unfortunately in India, the question has not yet received sufficient attention, though we must acknowledge with gratitude the excellent work done in this direction by some distinguished medical men and scientists foremost among whom may be mentioned Sir Robert McCarrison, late Director of Nutritional Research, in South India. And yet there is no country where the need for it is greater. The physical degeneration of the people of India has assumed alarming proportions, both among males and females. In the matter of infant mortality, India tops the list of all the nations of the world. In some towns it has approached 600 per thousand of the children born. The average of life in India is 22 years while that in Great Britain is 52 years.

This is a national calamity. India was the first country of the world to condemn the use of a diet which included flesh, fish, fowl and eggs. It was the first to discover the great advantages of the natural food of man—consisting of cereals and milk and fruits. The staple diet of the country as a whole is vegetarian, and the people have lived for ages upon it, in health and prosperity. But the impact of the meat-eating west has been undermining the faith of some of our people here and there, and they have been advocating the use of meat, and some of eggs, while the volume of opinion among scientists and food reformers in the west has been steadily growing against the use of meat. It is vital in national interest that this evil should be nipped in the bud, and that the existing knowledge of the Science of Nutrition which has been gathered in Europe and America should be widely disseminated in this country. For it is not merely the want of nourishing food, but also the lack of correct knowledge of what constitutes a well balanced diet, pure and cheap, and yet capable of giving strength, health and vigour, that is responsible for the present deplorable deterioration in the health of the people. Dr. Godbole has rendered a service to the country by writing this book, which he has done at my suggestion, as a means of propagating such knowledge.

The subject may be viewed under two main heads : (1) What not to eat and (2) What to eat. Under the former head Dr. Godbole has naturally discussed the question of vegetarianism and non-vegetarianism (Chapter IX), and of vegetarian *versus* non-vegetarian food (Chapter XVII). He has dealt with the question both from the scientific and the cultural points of view. The scientific objections to the use of flesh food have again and again been pointed out by a band of scientists and food reformers. But the results of scientific research require to be repeatedly proclaimed from the house tops until flesh food is abandoned.

The author tells his readers that "non-vegetarian food whether it be eggs, fish or flesh, is open to very serious hygienic objections, that it is positively harmful and leads to a multitude of diseases. And that in a country like India, where high-temperatures are common and where there is hardly any efficient supervision or legislation to control the production and distribution of eggs, fish and flesh, the danger accompanying non-vegetarian food cannot be exaggerated." (page 72). Some years ago Dr. Kellogg, the great American who has earned the gratitude of mankind and of the dumb creation by his powerful advocacy of a fleshless diet extending over more than sixty years, summed up his experience in the following passage : "After more than fifty years' personal experience with the fleshless diet and an unequalled opportunity to observe its effect upon both the well and the sick, the writer feels that he can say with full confidence that any reader who will make a careful study of the relation of meat-eating to intestinal toxemia or auto-intoxication, and will make even a brief trial of a well balanced and meatless bill of fare, will certainly be convinced of the immense advantages of the fleshless regimen and, in fact, of the impossibility of ridding one's self of the dangerous bacteria which infect the colon of every meat-eater and which give rise to colitis, appendicitis, and many other disorders (including headaches, high blood pressure, arteriosclerosis, and even Bright's disease), by any method other than the exclusion of fish, flesh and fowl from the bill of fare and with them the putrefactive bacteria which the butcher's meats always contain in countless millions."

Attention may also be drawn in this connection to the verdict of the International Scientific Food Commission which consisted of food scientists of four great nations, selected by their governments as experts, who decided in 1918 that no absolute physiological need exists for meat, since the proteins of meat can be replaced by other proteins of animal origin, such as those contained in milk, cheese and eggs, as well as by the proteins of vegetable origin. As Dr. Kellogg said years ago "the report of this, the most authoritative body which has ever spoken upon the food question, may be justly regarded as the last word upon the subject." The author further reinforces his argument by quoting the following important statements of Sir Robert McCarrison: "If the food contains enough milk and milk products then it is not necessary to eat flesh meat at all." "Enough milk is not less than a pint a day, and a quart if possible." "Eggs are not so good as milk." "One ounce of any of the 'dals' contains as much protein as one ounce of meat, nearly twice as much as one ounce of eggs, and seven times as much as one ounce of whole milk;" (page 76). It may be added, "that eggs, which undergo putrefaction as does meat also, can be eliminated from the bill of fare without injury, provided milk is eaten, and also provided care is taken to make a liberal serving of greens a part of every day's menu to ensure an adequate supply of food iron."

Dealing with the cultural and humanitarian point of view of flesh eating, Dr. Godbole well observes: "Man does not live for bread alone. As a representative of the highest evolution of rational beings, he has to live for some ideals. The loftiest of these ideals it cannot be denied is 'to live and let live'. This ideal is embodied in the great Sanskrit motto **"अहिंसा परमोधर्मः"**, 'not to injure the innocent is the highest form of religion.' It is for these reasons that the staunch vegetarian refuses to include in his menu anything like eggs, fish or meat. But the strict vegetarian stands on humanitarian ideals when he extends the conception of humanity from man to all creatures having life and feelings—the cow, the sheep, the pig and so on. The European conception of humanity strangely enough stops at human beings, and does not include the life of animals like the cow, sheep or the pig, (or the egg with life in embryo) which it thinks are outside the scope of humanity. The humanitarian in India looks upon the life of a cow that has yielded milk all her life time with the same feeling with which he looks at the life of one of his own relatives, and rightly too. If in addition to this, the question of food efficiency, cost and hygiene all help in preserving the ideal, there is no reason why this view should not be considered as the loftiest one to be fought for."

Dr. Kellogg has well pointed out the contrast between slaying to eat and eating vegetables. "When man slays to eat, what a picture arises, with knife and axe he turns upon his trusted friends,—the sheep that kissed his hand, the ox that ploughed his field. The air is filled with shrieks and moans, with cries of terror and despair: the soil is wet with warm blood, and strewn with corpses." Look now at the other picture: " When animal eats vegetable, there is no pain, no sorrow, no sadness, no robbery, no deprivation of happiness; no sun-light shut out from eyes that were made to see, no sweet melodies for ever shut away from ears that were made to hear, no simple delights denied to beings that God made, if not in his own image, at least so nearly like his image, man, that the man whose eyes have been enlightened by the study of nature may look down and see in the millions of beings that God has made to share with him the divine spirit, the breath of life, some traits of himself that must now and then bring blushes to his cheek or strike deep into his soul barbed arrows of remorse." A contemplation of this contrast should turn a man's mind away for ever from flesh, fish and fowl.

The next question—what to eat—is already answered in what has been said before. "The bill of fare which wise nature provides for man in forest and meadow, orchard and garden, a rich and varied menu, comprises more than 600 edible fruits, 100 cereals, 200 nuts and 300 vegetables—roots, stems, buds, leaves and flowers—to say nothing of milk and various other dairy products" (Kellogg). The author reminds the reader that for the normal growth and repair of the human body, and for generating the energy which is necessary to maintain its normal temperature, the normal food of man should include carbohydrates, fats and proteins, in certain definite proportions. By a diagram he shows the composition of milk, which contains carbohydrates, fats, proteins, mineral salts and vitamins, A. B. C. D. and E. He has put in a number of comparison tables showing the food value of milk as compared with that of different kinds of meat and eggs and of a large number of cereals, fruits, nuts and vegetables, and has reiterated the conclusion, well established by scientific research, that both for the young and the adult, Cow's milk is the most valuable—the most perfect—of all human foods, and at the same time the cheapest nourishing material which is available to man.

As the author has pointed out by references to the Vedas which are by common consent the oldest books extant in the world—and the Ayurveda, the Medical Science of the Hindus—the ancient sages of India were the first to recognise the high value of cow's milk both as a preventive and curative medicine and as nutritive food giving health, strength and long life, and marked the gratitude that man owed her by making it a matter of religion, of conscience and justice, that the life of a fellow creature, who served man as a most beneficent nurse should be protected by law. She is described in the Vedas as *Aghnaya* (one not to be killed), and held as an object of veneration. Excepting a few, in Indian States, the cow is still protected by law. But it is a great national misfortune that as the British people have been used to taking beef in their own cold country, since the establishment of British rule in

India, an appaling number of cows has been slaughtered in British India for supplying food primarily to the Europeans, in lamentable disregard of the deepest religious sentiments of the great bulk of the people, and of their economic and hygienic interests. The result of the consequent paucity and dearness of milk is writ large in the high infant mortality, the very low average expectation of life, and the increasing debility and decreasing vitality of the population of the country. Attention may be drawn here to the table (opp. p. 21) which shows the comparative consumption of milk and butter in India and other countries of the world. It has to be noted with regret that milk is many times dearer in Bombay than it is in London, Denmark and New Zealand. As the author says : "All the evidence available points to the conclusion that the consumption of fresh milk in India is very small when compared with such countries as the U. S. A., Denmark, Sweden and Switzerland. Since the desire for milk is wide spread and the consumption is relatively small, it would appear that the difficulties in the way of economic milk production and distribution in India are formidable." (page 23).

In India, attention has repeatedly been drawn during the last forty years to the milk problem. Sir Robert McCarrison wrote in 1928 in his excellent book on "Food":

"The greatest nutritional need of India at the present time is the production of more and of purer milk, for there is no more important food-stuff than this and none on which the public health is more dependent.............The rising generation must realise the importance of pure milk to the well being of the people, and see to it that India shall become as enlightened with regard to milk-production and to milk-distribution as other countries."

"If the food contains enough milk and milk products then it is not necessary to eat flesh meat at all. 'Enough' milk is not less than one pint a day, and a quart if possible; but few children in India consume anything like this amount."

As Prof. Kanga has pointed out in his valuable brochure on " Dietetics, Food and Race ", Sir Robert McCarrison has shown by his classical experiments on rats that among the national diets of India, the Sikh diet is the best. This consists of whole wheat chapaties, *urdki dal* (pulse), green vegetables in plenty, butter, fresh milk in plenty and "occasionally meat." In the U. P. and Rajputana, *Moong, Masur, Arhar* or *Chanaki dal* is used in the place of *Urad*. "If", says the professor, "the last item in the Sikh diet, *viz.*, "occasionally meat" is replaced by the Soya bean, it will be an ideal diet for the vegetarian; and as the Sikh diet has been found to be best among the national diets of India, and as Soya bean is undoubtedly superior to meat in its food value, the Sikh diet modified as suggested here would be certainly a more nutritive diet than the Sikh diet. The superiority of the Sikh dietary would be not only maintained but considerably enhanced by replacing meat by Soya bean."

The following questions and answers before the Royal Commission on Agriculture in India in 1926, clinch the matter so far as the importance of milk in the Sikh diet is concerned.

Q. 1479—Sir Henry Lawrence: Could you tell us a little more about this question of milk in diet? Do you regard a greater supply of milk as a very important matter for the rural population?

Health Commissioner (India): I think so. I have very little doubt that a population which has access to large quantities of milk is enormously benefitted thereby.

Q. 1480—Is there a geographical distribution? You say there are certain areas in India where they have sufficient milk and certain areas where they have not?

Health Commissioner: I do not think there is any doubt that much more milk is consumed in certain areas than in others, which means that the amount of milk is more readily available. In the Punjab, for example, the Sikh community drink, generally speaking, fairly large quantities of milk.

Q. 1481—With good effect on their health and constitution?

Health Commissioner: We have only to look at them to see that.

Q. 1484—Then, for people whose economic condition does not permit them to have a very generous diet, you would consider milk a very important substitute and any improvement in the milk supply as a matter of very great importance to India ?

Health Commissioner : A matter of the very greatest moment

It is a matter for deep regret that notwithstanding these emphatic statements by men competent to speak and entitled to be heard, little action has yet been taken in any province to improve the milk supply for the people. It is a matter for thankfulness that Lord Linlithgow has drawn public attention to the indisputable scientific fact that a liberal supply of milk is an essential constituent of diet for growing children and that good nourishment in the early years of life is essential to the building up of a strong constitution in after life. It is now equally well-established that milk is necessary to maintain health and strength and vigour throughout life. This is evident from the "Drink more milk" campaign that has been carried on in England and Europe and America for sometime past, where both official and private enterprises have been organised for distributing cheap and pure milk to the people. It is devoutly to be hoped that the interest which His Excellency the Viceroy is taking in this matter will lead to a well planned national endeavour, in which the British Government and the public and the Indian States, will co-operate to secure an adequate supply of pure and cheap milk to the people, and humane treatment to the cow and her progeny if not to all other innocent creatures of God.

The chapter on breeding or feeding deserves special notice because the Government and the public need reminding that feeding is as important as breeding. The Royal Commission on Agriculture in India said in their report in 1928: "The two important factors in cattle improvement are feeding and breeding. We place feeding first because no outstanding improvement in the way of breeding is possible till cattle can be better fed. The crux of the situation is the period of scarcity which in most, though not in all parts of the country, is the two or three months preceding the break of the south-west monsoon. Since it is the curtailment of uncultivated land as population has increased during the past century that is the most obvious cause of the present overstocking of village grounds, it is not surprising that many witnesses have advocated the extension of grazing land." Along with breeding it is necessary simultaneously to improve the economic condition of the farmer who would then be in a position to buy good cattle, feed it well, and maintain it in proper condition. Reserving of pastures for grazing, special provision for the supply and protection of stud bulls, putting an end to the slaughter of milch cattle, and last but not least introducing a system of collecting milk from the countryside and transporting it where necessary in dry ice vans, to cities like Bombay and Calcutta and other large centres of population and thereby putting a stop to the importation of cows for milking purposes into those centres, these are some of the immediate problems that must receive full attention before the question of an adequate milk supply can be solved.

The illustrations given in the book are happily designed and show orginality in the manner of their execution. They are meant to be at once educative, explanatory and convincing to the layman, whereas to the scientist they will be of more than common interest.

In concluding this note, I endorse the following observations of Prof. Kanga :

"If India is to become a strong and virile nation then it is the duty of Government, Municipalities, educational and other public bodies and the leaders and Panchayat of every community to see that no boy or girl leaves the school without getting an adequate knowledge of the important subject of dietetics and that no section of the public is without cheap and scientific dietaries suitable to its taste, purse and need."

M. M. MALAVIYA.

CONTENTS

— x —

INTRODUCTION.

During the course of a private coversation at the Vice-Chancellor's Lodge, one day, the subject of the food problem of India came up for discussion, particularly the question of milk and other vegetarian and non-vegetarian food stuffs. It was thought neccessary to find out the relative importance of non-vegetarian foods in Indian conditions and also to examine how far milk could be made to replace these. As the literature on this subject was far from complete, I was asked by our Vice-Chancellor Pt. Madan Mohan Malaviya to write a treatise on this subject and I took up the work immediately.

The main question was to examine the uses of milk as the most perfect food. All over the world, both in the West and in the East, there is a new wave of milk concious-ness. Nations that once believed only in maintaining liquor bars have now begun to encourage milk stalls. Country-wide campaign is being carried on in many Central European Countries demonstrating that both in cost and in food value, milk was any time superior to eggs, fish and meat. India has believed in the efficacy of milk as a perfect food for centuries past. The constant impact of western ideas has influenced a small section of the people to believe that non-vegetarian food was more efficient than milk but it is gratifying to note that milk has re-established its supreme place once again all the world over.

The dangers accompanying the use of non-vegetarian food stuffs in India are of a serious nature. As against the one advantage of easy assimilability of the nitrogen, the disadvantages are overwhelming. The abnormal heat of the tropical climate coupled with the want of strict sanitary control and cold-storage makes the use of non-vegetarian food a very risky one. There is no doubt that meat of all kinds, beef and pork particularly, are cheaper in India than in Europe or America, firstly because of the religious scruples (of the Hindus who would not touch beef and of the Mohamedans who would not take pork) and secondly because for want of control one does not know whether the meat on the market comes from a diseased, dead or a slaughtered animal. The fish and the egg are not open to less serious objections.

It is also worthwhile examining another serious aspect of this non-vegetarian question. This involves the consideration of the tremendous amount of vegetarian food that must be consumed by cows, buffaloes and goats etc., before it is assimilated and converted into a raw material for slaughter-houses. In the interests of national economy, would it not perhaps be more useful for man directly to use this vegetarian food in other forms rather than convert it into non-vegetarian food and then reuse it ?

The problem of milk suply in India is an extremely perplexing one. There is no other country in the world that worships the cow as a sacred animal. Indeed according to one orthodox view in Hindu India, the cow is the home of Thirty-three crores of Gods and Goddesses (making an average of one Deity per Indian!). In all provinces in India, even today, the cow is specially worshipped by womenfolk once a year in the evening, two days before Divali as a token of peace and prosperity. It is equally a paradox, however, that very little of pure and unadulterated milk is available in the country. The Indian agriculturist and the Indian milkman who look upon cow-slaughter as a religious crime do not seem to think that dilution of milk with water (or *vice versa*) is a still greater crime. The explanation for this mentality is to be found more in the economic conditions than in religious beliefs. Though not in intellectual pursuits, poverty is certainly a bar in the economic struggle for existence.

It would be a great mistake to interpret the religious importance attached to the cow as being a superstition. In India, religion is very often only a garb and a cover to disseminate cultural, sanitary, scientific and economic ideas. The value of the cow and the bullock to the Indian agriculturist cannot be minimised, the former for its milk and the latter as the beast of burden. The importance attached to the horse in the west as a valuable beast of burden both in agriculture and in war is a parallel conception. Though not on religious grounds, the flesh of the horse scarcely finds a place in western menu, except in times of a siege. The horse is as indispensable in the west as the cow and bullock are in the east. The religious significance attached to the cow in India is synonymous with the economic sanctity attached to the horse in the west.

Even in a country like Japan where the science of drinking milk was unknown, and people were afraid of taking it lest horns might grow on their heads, there is now a country-wide awakening in favour of milk and the production of milk, butter and condensed milk has reached appreciable figures. In the year 1930, for a total population of 64 millions in Japan, the production of milk amounted to 110 millions of litres, condensed milk amounted to 13 millions kilogrammes and butter amounted to 4 and a half millions kilogrammes. For a country that did not dare to take milk even in its tea, the above figures are certainly praiseworthy.

The following pages are written with a view to examine the question of milk from all aspects. No attempt is made to appeal to sentiment in preference to scientific reasoning. It is hoped that they will serve the purpose of disseminating the correct knowledge of milk as the most perfect food.

I

WEIGHTS AND MEASURES

In Europe, an average male person is taken to weigh 154 lbs. and an average female is taken to weigh 120 lbs.

Liquids.

One Imperial gallon (of water) = 10 lbs.

4 litres = 1 gallon.

1 litre = 1000 cubic centimetres.

or = 1000 millilitres.

One cub. centimetre = one millilitre.

1 c. c. = 16 minims

1 pint = 20 ozs. (fluid)

2 pints = 1 quart

4 quarts = 1 gallon = 8 pints.

30 c. c. = 1 oz. (fluid)

Solids.

1 gramme = 15·5 grains

1 lb. = 453 gms. = 7000 grains = 8 chataks

1 oz. = 28·3 gms = 2½ tolas.

1 Kg. (Kilogram.) = 2·2 lbs.

1 Seer (Bengal) = 2 lbs = 80 tolas.

1 Tola = 180 grains.

40 Tolas = 1 lb.

1 Chatak = 5 tolas.

16 Chataks = 1 seer.

Thermometry.

There are two scales used to measure heat in degrees. One is known as the Centigrade system used in all scientific books and also in France, Germany etc., but not in India or England. The other is known as the Fahrenheit scale used mostly in India and England.

The freezing point of pure water on the Centigrade scale is 0° C. and the boiling point of pure water at sea level is 100°C. On the Fahrenheit scale, the freezing point of pure water is 32°F., and the boiling point of water is marked 212°F. Therefore, 100 degrees of the Centigrade scale (100° – 0° = 100), correspond to 212° – 32° = 180 degrees on the Fahrenheit scale. The following simple formula is used for converting Centigrade into Fahrenheit and *vice versa.*

$\frac{F-32}{9}=\frac{C}{5}$ where F stands for degrees in Fahrenheit scale and C stands for degrees in Centigrade scale.

Indian Coinage.

		s. d.
One Rupee	=	1 6
One Rupee	=	16 Annas.
One Anna	=	4 Pice.
One Pice	=	3 Pies.

II

VEDIC AND AYURVEDIC REFERENCES

A. Milk in Vedic Times.

It is interesting and instructive to find that references of educative value should be found in the Vedas. The following selections will, therefore, be read with very great interest.

(1) गोभिः श्रीणीत मत्सरम् ॥

ऋग्वेद ६ । ४६ । ४

"Milk should be added to सोमरस."

This drink is recommended to be healthy.

(2) पुष्टिं पशूनां परिजग्रभाहं चतुष्पदां द्विपदां यच्च धान्यम् ।
पयः पशूनां रस ओषधीनां बृहस्पतिः सविता मे नियच्छात् ॥

अथर्व ॥ १६ । ३१ । ५

"I have obtained abundant wealth of cattle, bipeds and quadrupeds and corn in plenty.

May Savitra (God Sun) and Brihaspati vouchsafe me the milk of kine and the refreshing juices of herbs."

(3) वशाया दुग्धं पीत्वा साध्या वसवश्चये ।
ते वै ब्रध्नस्य विष्टपि पयो अस्या उपासते ॥

अथर्व वेद १० । १० । ३११

"When these Sadhyas and Vasus have drunk the outpourings of the cow, they pay adoration to her milk in the Bright One's dwelling place."

The whole of this Sukta, containing 34 Mantras is important and shows clearly the Vedic attitude towards cow and her milk.

(4) पयो धेनूनां रस ओषधीनां जवमर्वतां कवयो य इन्वथ।

अथर्व वेद ४ । २७ । ३

"Bards who invigorate the milk of milch-kine, the sap of growing plants, the speed of coursers."

(5) आहरामि गवां क्षीरं आहार्षं धान्यं रसम्॥

अथर्व २ । २६ । ५

"Hither, I bring the milk of co ws, hither brought the juice of corn."

(6) संसिंचामि गवां क्षीरं समाज्येन बलं रसम्।

अथर्व २ । २६ । ४

"I pour together milk of kine with butter blending strength and juice."

It is interesting to note that no reference is made in the "Vedas" to the milk of buffaloes.

B. Ayurvedic References.

Sushruta has given a very detailed description of the various properties of different milks and also those of milk in general and the following translation of the forty-seventh, `forty-eighth and forty-ninth Shlokas of the forty-fifth chapter will be found instructive :

" The milk is the white fluid essence of drugs and cereals, which enter into the food of the aforesaid milk-giving animals (cow, she-goat, she-camel, ewe, she-buffalo, mare, she-elephant or woman) and is therefore the best of all nutritive substances (literally life-giving). It is heavy, sweet, slimy, cold, glossy, emollient, laxative and mild.

" Hence it proves congenial to all sentient animals. And since milk is kindred in its nature to the essential principles of life and so very congenial to the panzoism of all created animals, its use may be unreservedly recommended to all and is not forbidden in diseases due to the deranged action of Vayu or Pittam or in ailments affecting the mind or the vascular system of man. Its beneficial and curative efficacy may be witnessed in cases of chronic fever, in cough, dyspepsia, phthisis and other wasting diseases, in Gulma (abdominal glands), insanity, ascites, epileptic fits, in vertigo, in delirium, in burning sensation of the body, in thirst, in diseases affecting the heart and the bladder, in chlorosis and dysentery, in piles, colic and obstinate constipation, in Grahani, Pravahika, miscarriage and other diseases peculiar to the female reproductive organs, and in haemoptysis. It is a refrigerant and acts as a bracing beverage after physical exercise. It is sacred, constructive, tonic, spermatopoietic, rejuvenating and aphrodisiac. It expands the intellectual capacities of a man, brings about the adhesion of broken or fractured bones, rejuvenates used and exhausted frames, forms an excellent enemata, increases the duration of life and acts as a vitaliser. It is an emetic and purgative remedy and imparts a healthy rotundity to the frame and which through its kindred or similar properties augments the quality of bodily albumen and is the most complete and wholesome diet for infants, old men and persons suffering from cachexia witnessed in cases of ulcers in the chest, as well as for persons debilitated from insufficient food, sexual excesses or excessive physica labour.

" Cow milk is demulcent, and does not set up or increase the normal quantity of slimy secretions in the internal channels of the body. It is heavy and is a good elixir and proves curative in haemoptysis. It is cold and sweet both in taste and chemical reaction. It subdues both Vayu and Pittam and is accordingly one of the most efficient of vitalising agents.

" The milk of a she-goat is possessed of properties similar to those of a cow and is specially beneficial to persons suffering from phthisis. It is light, astringent, appetising and is efficacious in dyspepsia, cough and haemoptysis. The milk of a she-goat proves curative in all diseases owing to the smallness of her limbs and her agile habits, as well as for the fact of her drinking comparatively a less quantity of water and living upon bitter and pungent herbs.

" The milk of a she-camel is parchifying, heating, light, palatable and possessed of a little saline taste. It proves curative in oedema, abdominal glands, ascites, piles, intestinal worms and Kushtha, and is a good antitoxic agent.

" The milk of an ewe is sweet, demulcent, heavy and proves aggravating in disorders of Pittam and Kapham. It forms a good diet in Kevala Vata and in cough due to the deranged condition of the bodily Vayu.

" The milk of a she-buffalo is sweet in taste, tends to impair digestion and increases the slimy secretion of the organs. It is heavy, soporific, cooling and contains more fatty matter than cow's milk.

" The milk of a she-animal with unbifurcated hoofs such as the mare etc., is tonic, light, parchifying, sweet and acid in taste leaving a saline aftertaste and proving curative in cases of rheumatism restricted to the extremities.

" The milk of a woman is cold and sweet leaving an astringent after-taste. It proves beneficial as an errhine and acts as a good wash in eye-diseases. It is wholesome, vitalising, light and appetising.

"The milk of a she-elephant is sweet though it leaves an astringent aftertaste. It is spermatopoietic, heavy, demulcent, cooling and tonic. It invigorates the eye-sight.

"The milk of a she-animal, milched in the morning is heavy, cold and takes a long time to be digested owing to her entire repose during the night when cooling attributes preponderate. Similarly the milk milched in the evening is found to be possessed of refrigerant and eye-invigorating properties. Moreover, it restores the bodily Vayu to its normal condition owing to the physical labour undergone by the animal in the day time, exposed to the rays of the sun and the currents of free air. Cold or unboiled milk is extremely heavy and serves to increase the slimy secretions of the organs whereas by boiling it is freed from those injurious traits. But this rule does not hold good in the case of woman's milk which is wholesome in its natural or unboiled state. Freshly milched warm milk should be regarded as extremely wholesome which being cooled down loses its efficacious virtues and becomes unwholesome. On the contrary over-cooked milk is heavy and fat-making. The milk which emits a fetid smell or has become discoloured and insipid or has acquired an acid taste and looks shreddy and curdled or tastes saline should be regarded as unwholesome and injurious."

(From an English translation of the *Sushruta* by Kaviraj Kunjalal Bhishagratna, vol. 1, 1907, pp. 430-434.)

A similar account is given in *Charak Samhita* Chapter 27, Shlokas 221-228.

Wagbhat (*Ashtang-Hridaya*) confirms the above in his Chapter V, Shlokas 20-29 and, while describing the individual properties of various milks, concludes by saying 'धारोष्णं अमृतोपमम्' *i. e.*, freshly milched milk is as nectar itself.

III

Imports of Milk Products, Butter, Ghee etc.

It is a great pity that a country like India with such a large number of cows and buffaloes and with a rich variety of climates and pastures should not only not produce enough milk and milk products to meet her own needs but should be compelled to import increasing amounts every year. The imports given in the following paragraphs point out the distressing drain on the slender resources of an impoverished country. The rate of the mortality of children in India, per thousand, is the highest on record. The reason is not far to seek. Many of the newly born babies do not get either enough or even a pure supply of natural milk and have to be fed on tinned substitutes. These tinned milks cannot obviously replace pure and fresh milk in all its food values.

For the vegetarian adult also, the supply of pure butter and ghee has been an equally tough problem. The high degree of adulteration combined with a lack of strict sanitary legislation has made the supply of pure butter and ghee a very intricate question. Cases are on record in the Bombay market where butter adulterated to the extent of as much as ninety eight per cent has been sold under the name of pure butter. The manufacture and import of vegetable ghee has given a very safe means in the hands of the ghee merchant of adulterating pure ghee. The import of a synthetic ghee aroma has futher simplified this adulteration.

The large amounts of synthetic food stuffs (containing milk) imported point to the increasing debility of the population of the country which year after year has been showing decreasing vitality. The statistics of the increasing imports of milk and milk products will, therefore, be found very instructive.

IMPORTS

STATISTICAL FIGURES SHOWING THE IMPORTS

OF

Milk, Condensed and Preserved, Milk Food for Infants and Invalids

CHEESE, BUTTER AND GHEE

I.—Milk foods for infants and invalids.

(Quantity in Cwts.)

	1930-31	31-32	32-33	33-34	34-35
Br. Emp.	12,718	11,272	9,735	11,541	8,999
Foreign cou.	660	572	582	486	175
Total ..	13,378	11,844	10,317	12,027	9,174

(Value in Rs.)

	1930-31	31-32	32-33	33-34	34-35
Br. Emp.	25,34,061	22,26,633	18,15,945	16,00,632	13,46,057
Foreign cou.	86,133	77,585	73,153	40,865	21,654
Total ..	26,20,194	23,04,218	18,89,098	16,41,497	13,67,711

*II.—Vegetable products (Ghee, fat etc.)

(Quantity in Cwts.)

	1930-31	31-32	32-33	33-34	34-35
Br. Emp.	782	641	484	211	183
Foreign cou.	2,95,239	1,16,208	36,175	2,438	4,589
Total ..	2,96,026	1,16,849	36,659	2,649	4,772

*The import of Vegetable Ghee has fallen because a number of hydrogenation factories have been started in India, in Bombay, Cawnpore, Lyallpur and Tatapuram.

(Value in Rs.)

	1930-31	31-32	32-33	33-34	34-35
Br. Emp.	29,630	26,745	19,777	8,297	7,031
Foreign cou.	1,08,73,769	42,14,892	13,25,948	94,871	1,28,893
Total ..	1,09,03,399	42,41,637	13,45,725	1,03,168	1,35,924

III.—Butter.

(Quantity in Cwts.)

	1930-31	31-32	32-33	33-34	34-35
Br. Emp.	2,273	3,166	3,556	4,618	5,767
Foreign cou.	352	404	216	488	498
Total ..	2,625	3,570	3,772	5,106	3,265

(Value in Rs.)

	1930-31	31-32	32-33	33-34	34-35
Br. Emp.	4,03,685	4,72,965	4,90,350	5,26,265	5,82,774
Foreign cou.	66,706	69,432	33,150	52,081	40,880
Total ..	4,70,391	5,42,397	5,23,500	5,78,346	6,23,654

IV.—Ghee.

(Quantity in Cwts.)

	1930-31	31-32	32-33	33-34	34-35
Br. Emp.	392	301	98	43	75
Foreign cou.	472	2,114	358	233	264
Total ..	864	2,415	456	276	339

(Value in Rs.)

	1930-31	31-32	32-33	33-34	34-35
Br. Emp.	25,357	16,925	6,320	2,750	3,660
Foreign cou.	29,655	1,14,632	18,739	11,871	11,251
Total ..	55,012	1,31,557	25,059	14,621	14,911

V.—Cheese.

(Quantity in Cwts.)

	1930-31	31-32	32-33	33-34	34-35
Br. Emp.	2,050	2,037	2,391	3,152	3,283
Foreign cou.	8,393	5,312	6,467	6,719	7,641
Total ..	10,443	7,349	8,858	9,871	10,924

(Value in Rs.)

	1930-31	31-32	32-33	33-34	34-35
Br. Emp.	2,35,822	1,80,832	2,27,666	2,76,608	2,72,710
Foreign cou.	7,69,622	4,52,625	5,33,485	5,30,650	5,70,682
Total ..	10,05,444	6,33,457	7,61,151	8,07,268	8,43,392

VI.—Milk, condensed and preserved.

(including milk-cream.) (Quantity in Cwts.)

	1930-31	31-32	32-33	33-34	34-35
Br. Emp.	54,612	30,416	41,024	57,733	62,244
Foreign cou.	1,72,241	1,55,509	1,31,308	1,14,137	1,18,698
Total ..	2,26,853	1,85,925	1,72,332	1,71,870	1,80,942

(Value in Rs.)

Br. Emp.	26,18,342	15,29,228	15,57,628	21,91,987	24,80,383
Foreign cou.	52,36,441	42,03,747	33,52,472	23,59,065	23,56,593
Total ..	78,54,783	57,32,702	49,10,100	45,51,052	48,36,976

Grand total for the year 1934-35.

	Rs.	
No. I.	13,67,711	Milk foods etc.
No. II.	1,35,924	Vegetable Ghee etc.
No. III.	6,23,654	Butter.
No. IV.	14,911	Ghee.
No. V.	8,43,392	Cheese.
No. VI.	48,36,976	Milk condensed and preserved.
Rs.	78,22,568	

Japan has been exporting to India a very large amount of hydrogenated fish-oil—both in the form of a fat and as fatty acids. The fatty acids are being used for soap manufacture and the fat is being used for adulterating ghee.

"Provisions—Vegetable products (Vegetable ghee etc.) increased from 3000 cwts. valued at Re. 1 lakh in '33-34 to 5000 cwts. valued at $1\frac{1}{8}$ lakhs in the year under review.

"Milk foods for infants and invalids imported mostly from the U. K. recorded a decrease in quantity from 12,000 cwts. to 9,000 cwts. and in value from Rs. 16 lakhs to Rs. 14 lakhs. Imports of condensed milk also rose from 1,72,000 cwts. valued at Rs. 45⅓ lakhs in 1933-34 to 1,81,000 cwts. valued at Rs. 48⅓ lakhs in 1934-35. Receipts during the year included 94,000 cwts. from the Netherlands and 56,000 cwts. from the United Kingdom as against 84,000 cwts. and 39,000 cwts. respectively sent by these countries in 1933-34. Other principal sources of supply were Italy 11,000 cwts., Denmark 10,000 cwts. and Australia 2,500 cwts.

Butter			Rs.
Import duty	1934-35	6,000 cwts.	6 lak hs.
25 %	1933-34	5,000 cwts.	6 lakhs.
ad valorem	1932-33	4,000 cwts.	5 lakhs.

Import duty on milk condensed, preserved and Cream....30 % ad valorem".

*Review of the Trade of India, 1934-35, p. 55.

IV

Total Production of Milk and Butter per year in India.

Perhaps the most difficult question in the milk problem of this country has been the attempt at a correct calculation of the total production and supply of milk and butter in the whole of India, including Indian States. Neither the provinces under the British Government nor the Indian States, small and large, have cared to keep a regular record of the milch cattle, cows and she-buffaloes, in their jurisdiction. Although attempts are being made for the last fifteen years to collect the necessary statistics regarding cows and she-buffaloes, a very vital question for an agricultural country like India, it is a pity that even in the year 1935, big provinces like Bengal and Behar and Orissa have not been able to collect this information "mainly owing to financial reasons" !

In the year, 1935, only sixty per cent of the Native States have been able to supply figures of all the cattle in their territories. It is hoped that at least in the next census of 1940, all these figures in British India and the Indian States will be as complete as the cause demands.

TOTAL PRODUCTION OF MILK AND BUTTER IN INDIA*.

Cows and Buffaloes in India.

ASCERTAINED BY CENSUS HELD IN 1930.

Provinces.	Cows.	She-buffaloes.
Madras	56,22,356	27,65,694
Bombay	27,36,924	15,04,066
Bengal	82,50,610	2,75,989

*INDIA :—Live Stock Statistics, 1929-1930. Report on the Third Census of Live Stock, Ploughs and Carts in India held in 1930; Department of Com. Intell. and Statistics, India.

Provinces.	Cows.	She-buffaloes.
Continued from Page 13.		
United Provinces	62,32,522	40,81,515
Punjab	24,18,444	27,27,264
Burma	14,67,349	4,05,740
Bihar and Orissa	57,92,528	16,25,792
C. P. and Berar	41,27,871	9,43,204
Assam	16,58,158	2,21,726
N. W. Frontier	2,77,398	1,48,715
Ajmer-Merwara	1,39,170	53,943
Coorg	39,040	6,838
Delhi	21,073	23,381
Manpur Pargana	2,019	670
British India Total ..	3,87,85,462	1,47,84,537

(In Indian States ascertained by Census held in 1930, pages 8 and 10 of the Report.)

States.	Cows	She-buffaloes.
Hydrabad	28,47,390	12,40,132
Mysore	15,96,909	4,82,951
Gwalior	9,76,766	4,52,569
C. I. States	15,50,590	6,22,876
Rajputana States ..	30,97,196	8,34,440
Madras ,, ..	5,40,011	54,813
U. P. ,, ..	3,07,624	1,30,736
Punjab ,, ..	5,78,771	4,42,422
Western Ind. ,, ..	6.34,781	4,21,733
Total ..	1,21,33,038	46,82,672

		Cows	She-buffaloes
Baroda, Kashmir, Bombay States and several States in Rajputana of which census was not taken.	..	15,00,000 (In round figures). out of total oxen* of 40,00,000	10,00,000 (In round figures). out of total of 2 millions of buffaloes.
Total, All Native States...		1,36,00,000 (In round figures).	57,00,000 (In round figures).

The figures relating to the first census refer to 28 States comprising about 22% of the total area of the Indian States, the second census to 40 States covering about 38 % of the total area, while those for the third census to 83 States, comprising about 50% of the total area. The census was not taken in Baroda, Kashmir, Bombay States and several States in Rajputana and Central India. Adding the latest available statistics of live stocks for those States, the total number of live stock, ploughs and carts, in Indian States, for which statistics are available, may be roughly computed as follows (all Native States together).

All Buffaloes	..	11 millions.
Oxen ..	..	38 ,,

Total number of milk-yielding cattle (cows and buffaloes) in the whole of British India and Native States, according to the census in 1930.

		Cows.	She-buffaloes.
Total British India	..	3,87,85,462	1,47,84,537
Total Native States	..	1,36,00,000	57,00,000
Grand Total	...	5,23,85,462 52·5 millions cows (in round figures).	2,04,84,537 20·5 millions she-buffaloes (in round figures)

Therefore, the total of Cows and She-buffaloes =

$$52{\cdot}5 + 20{\cdot}5 = 73 \text{ millions.}$$

**Note.*—Buffaloes include male-buffalo, cow or she-buff. and young stock. Oxen include bulls, bullocks, cows and young stock.

Total quantity of milk in lbs. as estimated from the figures of the Census of cattle for the year 1930 :—

The total number of cows in 1930 amounted to 52·5 millions and the total number of she-buffaloes amounted to 20·5 millions, in the whole of India. Making allowance for the dry period of cows and she-buffaloes and the amount of milk given to the calves etc., it is roughly estimated that

per cow	100 lbs. per annum
and per she-buffalo	1200 ,, ,, ,,

are available in India for the total population, (350 millions).

According to this approximate estimate, the total amount of milk available (both cow's and she-buffalo's) is

$$\text{lbs.}\ \frac{5{,}25{,}00{,}000 \times 100 + 2{,}05{,}00{,}000 \times 1200}{35{,}00{,}00{,}000}$$

$$= \frac{29{,}85{,}00{,}00{,}000,}{35{,}00{,}00{,}000} = \frac{2{,}985}{35}$$

85·3 lbs. per annum per head. *i. e.*, per head, per year about $8\frac{1}{2}$ gallons. It is so very difficult to hazard an average estimate for the different provinces of India.

" The figures for the year 1930 show that in Lahore, for a population of four lakhs, the total amount of milk consumed per day amounts to 1207 maunds, *i.e.*, per head, per day, it comes to about *2 Chataks* or *4 Ounces*. The average price in Lahore in 1930 is taken at As. 4 per seer of milk. According to the figures of the Census of the same year, in the Punjab, there were 26 lakhs of cows and 31 lakhs of she-buffaloes out of a total of 141 lakhs of cattle."

Up-to-date statistics are now available regarding the number of milch-cattle and the total quantity of milk procurable in the Bombay Presidency. The following extracts should prove a very interesting reading. "If all the milk produced in the Bombay Presidency was made available for drinking purposes only, the consumption per head of the population would be *worth less than one pice per day.*" The report of the Marketing Officer of the Bombay Presidency throws a flood of light on the industry

Note.—Average yield of butter fat from cows in New Zealand comes to about 240 lbs. per year and in Denmark it is about 300 lbs. per year.

and some of the statements and figures given are worth consideration. "The latest census of cattle in the Bombay Presidency shows that there are 11,75,354 buffaloes and 17,76,589 cows in the Presidency in addition to huge flocks of goats and sheep from which considerable supplies of milk are also obtained. It is rather an extraordinary fact that throughout the Presidency less than 2500 families maintain for their private use, either a buffalo or a cow, and of these few people, a big majority prefer to maintain a buffalo."

"It has been ascertained by research workers on the milk production problem, that a total of 98,69,13,430 pounds of buffalo's milk is produced (annually) in the Bombay Presidency as against 2,56,88,456 pounds of cow's milk. Taking an average of prices in all the districts in the Presidency, it has been found that buffalo's milk sells at approximately 1½ annas per lb., as against cow's milk at 1 anna per lb. The total value of milk produced from all sources may be taken as approximately :

	Rs.
Buffalo's Milk	9,25,23,134
Cow's Milk	1,60,55,285
Sheep and Goat's Milk	42,37,490
Total ..	11,28,15,909

"In the Deccan, cows are rarely maintained for milk and probably, on an average, do not yield anything more than *two pounds of milk a day.*" "In Poona, 12 per cent. of the total milk is used for tea, 20 per cent. for drinking purposes, 15½ per cent. for making into curds, 12½ per cent. for ghee and 40 per cent. is exported from the district."

There has been a serious fall in the total number of cows and she-buffaloes in the Bombay Presidency as the following statistics for 1930 and 1935 would show :—

	Cows.	She-buffaloes.
1930	27,36,924	15,04,066.
1935	20,28,729	12,70,311.
Fall in five years ..	7,08,195	3,34,755.

These figures make a very distressing reading !

Province.	1930	1935	1930	1935.
	Cows.	Cows.	She-buffaloes.	She-buffaloes.
Ajmere-Merwara ..	1,39,170	1,35,003	53,943	59,349
Assam	16,58,158	17,73,961	2,21,726	1,99,206
Bombay	27,36,924	20,28,729	15,04,066	12,70,311
Burma	14,67,349	14,18,393	4,05,740	3,95,301
C. P. & Berar ..	41,27,871	34,31,505	9,43,204	8,80,217
Coorg	39,040	35,195	6,838	6,622
Delhi	21,073	25,637	23,381	31,099
Madras	56,22,356	59,84,793	27,65,694	28,88,678
N. W. F. Pr. ..	2,77,398	2,46,651	1,48,715	1,44,386
Punjab	24,18,444	26,53,056	27,27,264	30,52,866
Sind	..	8,91,715	..	3,60,906
U. P.	62,32,522	65,05,347	40,81,515	42,45,006

N. B.—The figures for the Census of 1935 for Bengal and Behar and Orissa are not available.

Indian States.

State.	Census of 1930; 50 %Area.		Census of 1935; 66%Area.	
	Cows.	She-buffaloes.	Cows.	She-buffaloes.
Baroda	Not available.		2,09,892	3,38,882
C. I. States ..	15,50,590	6,22,876	20,14,368	7,12,308
Gwalior	9,76,766	4,52,569	9,84,445	4,64,821
Hyderabad ..	28,47,390	12,40,132	29,98,588	13,15,185
Kashmir	Not available.		6,67,487	3,35,570
Madras States ..	5,40,011	54,813	5,76,902	60,379
Mysore	15,96,909	4,82,951	15,70,852	5,03,162
Punjab States ..	5,78,771	4,42,422	8,88,209	5,77,946
Rajputana States ..	30,97,196	8,34,440	35,39,587	12,31,365
U. P. States ..	3,07,624	1,30,736	1,72,212	1,27,039
W. I. States ..	6,34,781	4,21,733	6,25,117	4,10,024
Grant Total ..	1,21,33,038	46,82,672	1,42,47,659	60,76,681

British India (excluding Bengal and Behar and Orissa).

	1930.	1935.	Increase or Decrease.	Per Cent.
Cows	2,47,42,324	2,46,29,985	Minus 1,12,339	−0·5%
She-buffaloes ..	1,28,82,756	1,35,33,947	Plus 6,51,191	+5·1%

Indian States.

	50% Area only.	66% Area only.
Cows	1,21,33,038	1,42,47,659
She-buffalloes ..	46,82,672	60,76,681

Census of 1935.

			She-buffaloes.
Cows	2,46,29,985	(Excluding Bengal and Behar and Orissa)	1,35,33,947
Cows	1,42,47,659	(Native States 66% Area)	60,76,681
Total ..	3,88,77,644	Total ..	1,96,10,628

V

Comparative Use and Consumption of Milk and Butter in India and in other Countries.

A very serious difficulty in the calculation of the total annual supply of milk in India has arisen, on account of the different breeds of cows and buffaloes in the country and the varying quantities of milk they yield per day. Many of the agricultural experts are afraid of hazarding an average figure for the yield of milk per cow or she-buffalo because the daily yield of milk ranges from as low a figure as two pounds in the case of a cow per day to about sixty pounds per day in the case of a she-buffalo in another province. In the following calculation, therefore, a very safe average has been taken and it is hoped that the estimates are correct.

It will be seen from the following tables that, of all nations, India stands at the lowest rung of the ladder of nutrition and in spite of her increasing population, the number of milch cattle is either stationary or is even decreasing. It is a great pity and a disgrace that a small country like Denmark (of half the size of Mysore State in India) should export one hundred and seventy two millions of Kilos of butter, in addition to cream and eggs etc. and India should import four thousand cwts. of butter annually. What influence all this can have on the health of the nation can easily be understood.

COMPARATIVE CONSUMPTION

OF

Milk and Butter

(Used in the Country.)

	Milk (Gallons).	Butter (lbs.)
Finland	83·8	20·7
*Switzerland	70·4	13·0
Sweden	69·7	18·6
Norway	56·0	9·6
America	55·3	17·3
Canada	51·0	29·3
Czechoslovakia ..	45·8	9·9
Austria	45·0	3·3
*Netherlands	42·7	12·6
*New Zealand	37·4	34·1
*Australia	37·1	29·8
England	30·9	16·0
Germany	27·3	19·7
France	25·0	8·5
*Denmark	22·0	12·2
India	8·5	..

[*These countries export a large amount of Butter, Condensed Milk etc., in addition to what they use in the country.]

*The climate makes the keeping and transport of fresh milk difficult. There are no statistics on the subject; but it would appear that the greater part of the milk produced is consumed in the form of 'ghee', curds, and sweetmeats. In all the larger towns, the supply of fresh (liquid) milk is small: in Bombay it has been estimated at about 7 gallons and in Calcutta, at about 8 gallons per head per annum, *i. e.*, less than 3 ozs. per head, per day. The price of pure milk is high and if it was reduced to half the current rates, there is little doubt that the consumption would be more than doubled. Owing to an increase in the habit of tea-drinking in recent years, the demand for milk in urban centres has increased and there are now considerable imports of condensed milk especially into Burma.

Throughout India, generally, the supply of fresh milk in villages is said to be defective. In the C. P. the supply is estimated at less than 3 oz. per head daily—or 6·5 gallons per head per annum; in Bombay, most villages are short of milk; in Madras and in the U. P. the supply of fresh milk in villages is said to suffice for the demand. In Bihar and Orissa, the supply is considered to be too small for the cultivator's house-hold needs.

All the evidence available points to the conclusion that the consumption of fresh milk in India is very small when compared with such countries as the U. S. A., Denmark, Sweden and Switzerland. Since the desire for milk is wide-spread and the consumption is relatively small, it would appear that the difficulties in the way of economic milk production and distribution in India are formidable.

---

*Extract from the Report of the Royal Commission on Agriculture.

VI

AVERAGE DAILY WORK, FOOD AND CALORIES.

Work, Food and Calories.

One of the greatest difficulties in India for determining the quantity of food required per head—man, woman or child—is the want of data on average work done per head per day. In European countries, scientific experiments have been performed and the necessary data giving the relation between the amount of average work done and the food required has been fixed. In India, on the other hand, not only is very little scientific work done on the subject but there has been a lot of confusion even in blindly copying the foreign data and applying them to Indian conditions. For example, more than one scientific worker* has taken it for granted that in India, an average adult requires about 3000 calories per day, an average meant for European workers. Another author† of a pamphlet on food makes a similar supposition and takes it for granted that "An Indian man in the prime of life requires each day from 2500 to 3500 calories according to the part of India in which he lives and to the work he has to do; an Indian woman requires four-fifth of this amount, or from 2100 to 2900 calories". This same author, while describing that a resident in Northern India (*e.g.*, Punjab or Kashmir) would require more food than a resident in South India (*e. g.*, Madras), has overlooked the fact that even in Northern India (*e.g.*, Punjab) the same individual should require different quantities of food in winter and in summer, because whereas in winter the minimum reached may come to 32°F. (body-temperature being 98·5°F.) the maximum reached in summer may go up to 120°-125° F. (body-temperature being 98·5°F.). With these extreme differences, it is but natural that even in the same Northern India, the amount of food required per head must vary widely and with the same individual. It would there-

*Dr. N. R. Dhar, in an article in the *Leader*.
†Col. Robert MacCarrison, *Food*, pp. 110-11.

Body Temperature and Surrounding Temperature.

IN

India and Central Europe.

Body Temperature, in India, is *lower* than the surrounding temperature in summer and is *higher* than the surrounding temperature in winter..

In Central Europe, the body temperature is *higher* than the surrounding temperatures all the year round.

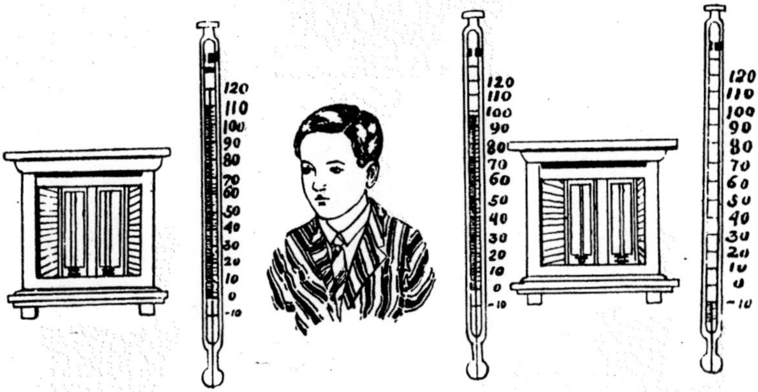

Body Temp.

98·4°F.

INDIA

Max. +120°F.

Min. + 35°F.

CENTRAL EUROPE

Max. +90°F.

Min. – 20°F.

HEAT ENERGIES OF FATS, PROTEINS AND CARBOHYDRATES.

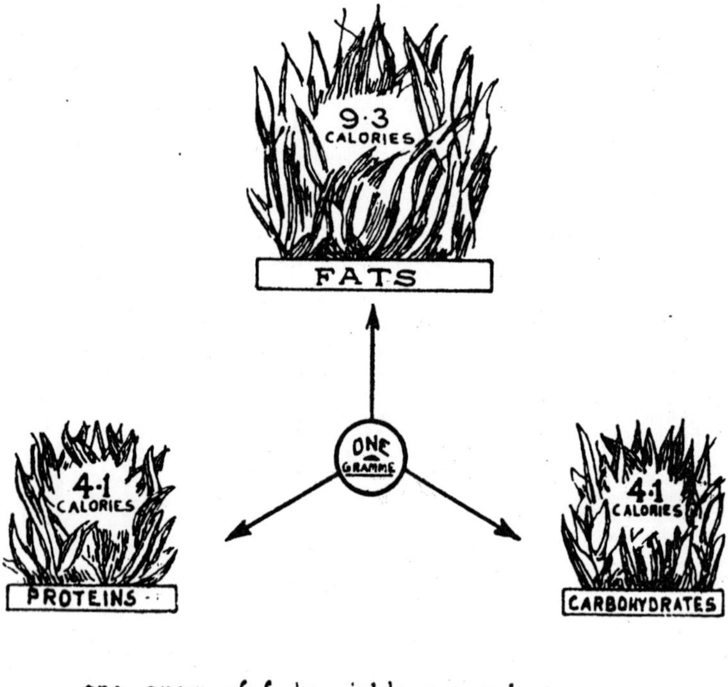

one gram of fats yields 9·3 calories.

one gram of Proteins yields 4·1 Calories.

one gram of Carbohydrates yields 4·1 Colories.

fore be necessary to examine this question from all points of view.

According to the principle of Conservation of Energy, energy can be neither created nor destroyed, and all work that is done can be expressed in terms of energy. Whenever the human body does any work, internal changes are taking place and an exchange in energy is brought about. The energy used must be replaced and this is done by the food material that is supplied to the body. This exchange which is taking place constantly, whether the body is at work or not, is dependent upon certain influences such as the temperature of the surrounding atmosphere (compared with the body-temperature), surrounding humidity, movements of the air, sun-shine, baths, dresses, etc.; not merely these but it also depends upon the largeness or the smallness of the same animal. As a rule, taken per unit of the surface of the body, smaller animals require greater exchange of energy than larger animals of the same species. This unequal exchange is dependent upon the relation between the surface and the mass of the body. The smaller the mass the greater is the surface in comparison with the mass. For example, if a cube of one-inch side is taken, the total volume is one cubic inch, whereas the surface is six square inches, *i. e.*, the relation between the mass and the surface is as 1:6. If, however, the side of the cube is two inches, the total capacity is eight cubic inches, each square surface measures four square inches and the total surface measures twenty-four square inches. In this case the relation between the mass and the surface is as 8:24 or as 1:3. That is to say as the surface increases the mass comparatively decreases. What is to be noted is that it is the surface that determines the ratio of radiation and that is the determining factor in deciding the loss of energy to the surrounding medium. It is easy to understand, therefore, that the smaller animals require more energy than the larger animals (of the same species) when every thing is calculated on the basis of one unit of the weight of the body. Based upon this argument, it has been calculated, under European conditions, that on the unit of one kilogramme weight per day, the exchange of heat in calories in the case of a strong man is 37 calories, in the case of a child in the first month 90 calories, and in the case of ten-year old child 60 calories and a dwarf (one of imperfect growth) 80 calories. Calculated on one square metre of the body surface, the corresponding energy exchange is as follows: in the case of a strong man, 1390 calories, in the case of a baby-in-arms, 1226 calories, in the case of a ten-year child, 1390 calories and in the case of a dwarf 1231.

The greatest influence exerted on this exchange in the body is dependent upon the total amount of work done.

By one "big calory" is understood the amount of heat energy required to raise one litre of water by one degree centigrade; expressed in terms of work this is equivalent to the amount of work that is done in raising a weight of 425 Kg. (11·5 maunds) through a height of one metre (about 3 feet 3 inches). In Europe, the amount of work done by an able-bodied worker, working for eight hours a day, is supposed to vary from 2,00,000 to 3,00,000 metre kilogrammes, *i. e.*, calculated in maunds, it ranges from the amount of work done in raising a weight of (11·5 maunds $\times$ 470·5) through a height of one metre or 3 feet three inches to (11·5 $\times$ 1·5 $\times$ 470·5) maunds through a height of 3 feet 3 inches. It is also calculated that in Europe, by merely moving about for eight hours without doing any hand-work, nearly 40,000 metre kilogrammes of work are done in a day, *i. e.*, about 11·5 $\times$ 100 $\times$ 39 maund-inches.

It is a well-known principle in thermo-dynamics that although the amount of work done can be expressed quantitatively in terms of energy yet the converse does not hold good. For example, all the coal that is spent in an engine is not quantitatively converted into work, *i. e.*, all the available energy is not transformed into work but only a part of it. In the case of steam engines, for example, hardly 10-15% of the coal is converted into work. The human body is also a machine and is, therefore, no exception to the law, and for all practical calculations nearly four to five times the amount of nourishing material has got to be used relatively to what corresponds to the actual amount of work done, theoretically.

Based upon these calculations involving important factors such as the temperature of the body, the surrounding temperature, the amount of humidity, the nature of the dress used, the minimum amount of work done either in a factory of eight working hours a day or by the average layman doing very little manual work and the time spent in sport etc., and the average physique of the worker, it is estimated that for the average individual in Europe, between 2000-3000 calories, *i. e.*, an average of 2500 calories are required per day. The following is a table quoted from a well-known German author:

Work.	Calories.	Proteins. *gms.*	Fat *gms.*	Carbo-hydrates.
A strong man working very hard ..	4800	133	100-150	500-600
A strong man doing average work..	3000	122	75-100	400-500
A strong man doing office-type work	2400	105	50	400-500
A weak man doing hard work ..	2400	75	60	400-500
A weak man doing ordinary work..	1800	75	40	300-400
An old woman resting	1600	60	30	250

From the above considerations it would be now easy o understand and appreciate the fallacy of taking the data for the average European and applying the same to the average Indian between whom very little is common.

In the first instance, the average life of the Indian hardly extends over a span of 30 years, whereas his European colleague claims to begin to be young at the age of 45 to 50. An average Indian is hardly given to any kind of regular outdoor sport in his every-day programme of life. It is doubtful as to how many Indians do the amount of daily work attributed to the average European working eight hours a day, *i. e.*, 2,00,000 to 3,00,000 metre-kilogrammes as given above. Next, think of the extreme differences of climate between Europe and India. Excepting Kashmir over the rest of India, the summer conditions of Europe are even colder than the winter conditions of India, *i. e.*, temperatures ranging from minus 5° to minus 30°C are common all over Europe, whereas a temperature of 15°-20°C would be considered as very warm in Europe. In India, on the

N. B.—An average man (in Europe) is taken to weigh 154 lbs. and an average woman weighs 120 lbs.

other hand, excepting Kashmir, scarcely a temperature of 0°C (32°F.) is reached even for a few hours during the 24 hours of the day. If a careful study of the winter conditions is made all over India, it will be seen that in presidencies like Bengal, Madras, Bombay, Central Provinces etc., the extreme conditions of European summer are scarcely experienced even in an average winter. Off and on, a few cold waves are known to pass over certain parts of India but they are known to last only for a few hours and that too only during a few nights. Remembering that the average body-temperature is about the same all over the world, the maintenance of the body-temperature over the surrounding temperature needs a very large amount of heat for radiation. A difference of 45°-60°C between the body-temperature and the surrounding winter temperature extended over a period of some months in winter is not an uncommon phenomenon in Europe, whereas under Indian conditions, the amount of heat required for the corresponding radiation is very much smaller. Added to this, it should not be forgotten that except in rainy season, the amount of the warmth supplied by the sun is a tremendous factor in favour of the Indian. More than half of the Indian population can move about half-naked during the day time, even in winter, thanks to the warmth supplied by the sun.

Then, again, think of the conditions in the Indian summer, particularly in areas like the Punjab, Rajputana, Sind, United Provinces, Central India etc. Some authors have indulged in the fallacy of speaking of the need of food in North India as though there was only one kind of average climate all through the year. Think of the minimum of 30°F of a wintry night at a place like Lahore and compare it with a maximum of a temperature of 120°F on a summer day at the same place. And it should not be forgotten that the summer extends over several months. It is obvious, therefore, that the needs of winter and summer are different even in the same North India. It would be interesting to find out what kind of food and how much of it should be administered to the average Indian in India in one of these summers when the body-temperature has to be kept at 98°—98.5°F while the surrounding temperature ranges between 100°—120° F.—by no means an uncommon phenomenon. By the laws governing temperatures, it would be necessary to cool down the body by helping it to evaporate moisture through the pores of the body to keep the temperature of the body below that of the surrounding medium. It appears that to think of loading the body with food enough to generate 2000-3000 calories per day when the body can hardly do any manual work, would be almost a tragedy.

The monsoons in India are another important feature which should not be forgotten in discussing the problem of food. Whereas in summer the distress is due to the high temperatures, low humidity and hot breezes, in monsoons, there is a greater distress due to lower temperatures, oppressive humidity and no breezes. The problem of food supply to the body under these latter conditions also needs careful investigation. Even Europeans or European settlers in India have been convinced that they are incapable of doing even half the amount of the work they do in Europe, when they are under these Indian conditions.

In addition to the considerations of temperature, radiation and average daily work etc., there is also the important question of the average weight of the Indian which must influence the total work done and hence the daily need of the calories. In Europe, the average male adult is taken to weigh 154 pounds and an average female is taken to weigh 120 pounds. In India, no such information is available from any responsible source. From what one knows of the general health of the country, it is doubtful whether the average weight of the male adult in India today is equal to the average weight of the female in Europe ! The deterioration in the youth of the country is so pitiable in the last twenty five years. This is, therefore, a serious consideration that goes against the unscientific application of the European data to Indian conditions.

In India, everyone is tempted to bathe or rather remain immersed in cold water in the summer months in the typically hot districts to keep down the temperature of the body. This is an operation which is the reverse of supplying food to the human body.

All these considerations are sufficient to convince anybody that it is a great mistake to copy the European data regarding food, dress and work and to apply them wholesale to Indian conditions. Thus, one feels a genuine pity for those who would be the victims of the general prevailing habit in India of wearing stiff and uncomfortable European dress in all sorts of climes and weathers.

For a proportionate and healthy growth of the human body, normal food should consist of (*a*) carbohydrates, (*b*) fats and (*c*) proteins in certain definite proportions. At least 12 % of the enengy required should come from proteins, *i. e.*, to say, for European conditions about 360 calories, (out of the average 3000 cal.) should come from proteins which would mean an average supply of 90 gms. of protien for the average European. Salts, vitamins and spices are also indispensable to activate the whole growth. Spices help digestion besides increasing the appetite. One very rich source of salts is the vegetables which also act as "excitants". The total heat required by the human body for being converted into work not only varies according to the average work done by an individual but also depends upon the difference between the temperature of the surrounding atmosphere and the normal temperature of the human body (as discussed above in detail).

According to accepted physiological laws, the amount of food taken is meant to serve two purposes; firstly to maintain the normal temperature of the body and secondly to replace the constituents of the body used up in doing actual work. So far as Indian conditions are concerned, it is perfectly reasonable to suppose that during the summer months as well as in the rainy season, when the surrounding temperature is high, not much heat energy is required for radiation to keep the body-temperature normal. It is, of course, necessary to take a certain amount of food to replace what the body is losing in doing the day's work, both physical and mental. The calories of heat required for the daily work have got to be distributed between the carbohydrates, fats and albuminoids in the ratio of about roughly 5:1:1, all of which together supply the necessary amount of energy required by an adult. In India, unfortunately, in the different provinces, depending upon the crop that is grown in the province, more of one particular type of food is taken than a general mixture so as to supply the albuminoids, fats and carbohydrates etc., in their proper proportion. For example, in the northern parts more wheat is taken than rice both in winter and summer. In summer, in the north, it is worthwhile considering whether wheat should not be substitued by something lighter when the high temperature conditions and the work done during the day need a very great modification. In Madras, Bengal and the west coast on the other hand, too much of rice is taken all the year round. The correct thing to be done is a combination of the different food-stuffs to suit the above ratio. It is worthwhile finding out how much and what quantity of food should be useful to a worker who taxes his brain more than his body. As is well-known, lecithin, a derivative of glycerides containing nitrogen and phosphorus is especially found in the brain and the nerves. Food containing lecithin, for example, eggs (for the non-vegetarians), almonds, soya-beans and some other leguminous beans (for the vegetarians), should be valuable for the brain-worker. From this point of view, milk is a very valuable food because it contains all the necessary ingredients both for the invalids as well as for the adults, particularly so for the former. An examination of the various types of milk from different animals will, therefore, be of very great interest.

VII

Composition of the different types of Milk and their nutritive values.

The composition of the milk of different animals forms a very interesting study both from the chemical and physiological points of view. It should be borne in mind that the object of nature is to feed the young ones of different animals with a necessary food stuff that will supply the different constituents necessary to build up the body of that particular species in a fixed time. The composition of the milk of the elephant is, therefore, different from that of the cow or the rabbit. If in one case, the proportion of fat preponderates, in another, the protein may be in excess. Man has, however, used all his intelligence in diverting all these milk supplies for his selfish and humanitarian (!) purpose. The question before him is not how he could be of service to the young one of the species in its growth, but he is anxious to find how best he could use the milk for his own interest.

In the first instance, the analysis of a milk helps in the correct understanding of the constituents that go to make up the milk. Indirectly, however, it enables the scientist to transform one kind of milk into another by suitable additions and alterations of the ingredients. With the growth in the modern researches, new methods are being found out to detect even traces of metals in different milks and who knows, it may be that these very traces of metals may be extremely necessary and valuable for the growth of man's life. An attempt has been made to make the accompanying tables both complete and uptodate.

Composition of various milks (according to Godbole-Sadgopal) from different animals.

Serial number.	Origin.	Total solids.	Albuminoids.	Fat.	Milk-sugar.	Ash.†	Sp. Gravity.
*1	Cow (average of about 50 samples.)	12·9-14·5	3·4-4·0	3·0-3·85	4·5-5·2	0·45-0·65†	1·028-1·035
2	Goat (average of 23 samples.) ..	12·6-13·2	3·6-6·2	3·2-3·95	4·0-5·3	0·6-0·82	1·03-1·036
3	Sheep (average of 13 samples.) ..	15·5-19·5	5·8-7·5	5·2-8·6	4·5-5·0	0·6-1·3	1·035-1·042
4	Buffalo (average of 63 samples.) ..	18·0-22·5	5·3-6·15	6·5-8·75	5·0-5·4	0·7-0·95	1·038-1·042
5	Mare (average of 6 samples.) ..	9·5-11·2	2·1-2·55	0·6-1·8	6-8·5	0·3-0·4	1·03-1·036
6	Ass (average of 7 samples.) ..	9·16-9·53	1·6-2·0	1·3-1·5	6·28-6·8	0·4-0·48	1·023-1·035
7	Mother (average of 11 samples.) ..	11·5-13·5	1·0-1·65	2·0-3·65	5·8-6·5	0·15-0·25	1·03-1·034
8	Elephant (average of 2 samples.) ..	20·0-28·9	10·3-13·4	12·5-15·6	7·2-10·3	1·2-2·7	1·2-1·75
9	Sow (1 sample.)	21·7	8·6	9·78	6·5	1·9	1·403
10	Bitch (1 sample.)	14·6	4·7	4·2	3·8	0·52	1·038

†Regarding the ash-content of cow's milk, due to its estimation at a very high temperature, the values obtained are comparatively low, perhaps on account of the loss of sodium chloride and other volatile salts. Generally the average is taken to be about 0·78%

N. B.—According to recent researches, cow's milk is found to contain 0·25% citric. acid, other combined nitrogenous matter 0·26% phosphatide, 0·05%, enzymes. vitamines and sterine.

Composition of different Milks (Teichert.)

	Sp. Gr.	Tot.	Sol.	Water.	Fat.	Protein.	Milk sugar.	Ash.
Human ..	1·028	11·2	..	88·8	3·1	1·6	6·3	0·2
Mare ...	1·031	9·3	..	90·7	1·2	2·0	5·7	0·4
Ass ..	1·033	10·3	..	89·7	1·5	2·4	6·0	0·4
Bitch ..	1·045	23·0	..	77·0	9·3	9·7	3·1	0·9
Cow ..	1·032	12·5	..	87·5	3·5	3·5	4·7	0·8
Sheep ..	1·036	17·0	..	83·0	5·3	6·3	4·6	0·8
Goat ..	1·032	14·2	..	85·8	4·5	5·0	4·0	0·7
Reindeer ..	1·047	30·2	..	69·8	16·0	9·7	3·7	1·5

	Sp. Gr.	Water.	Protein.	Fat.	Sugar.	Ash.	Total.	Sol.
Woman ..	1·0298	87·58	2·01	3·74	6·37	0·30	12·42	..
Cow ..	1·0313	87·27	3·39	3·68	4·94	0·72	12·73	..
Ass ..	1·0320	90·12	1·85	1·37	6·19	0·47	9·88	..
Sheep ..	1·0355	83·57	5·15	6·18	4·17	0·93	16·43	..
Goat ..	1·0305	86·88	3·76	4·27	4·64	0·85	13·12	..
Mare ..	1·0347	90·58	2·05	1·14	5·87	0·36	9·42	..
Sow ..	1·0380	83·49	7·23	4·55	3·23	1·05	16·06	..
Bitch ..	1·0350	75·44	11·17	9·57	3·09	0·73	24·56	..
Elephant ..	1·0313	79·30	2·51	9·10	8·59	0·50	20·70	..
Camel ..	1·0420	86·57	4·00	3·07	5·59	0·77	13·43	..
Buff. (Egypt)	1·0350	82·25	5·05	7·51	4·44	0·75	17·75	..
Buff. Ind...	1·0298	81·92	4·25	7·55	4·75	0·89	18·08	..
Reindeer ..	1·047	69·8	9·7	16·0	3·0	1·5	30·2	..

COMPOSITION OF MILK

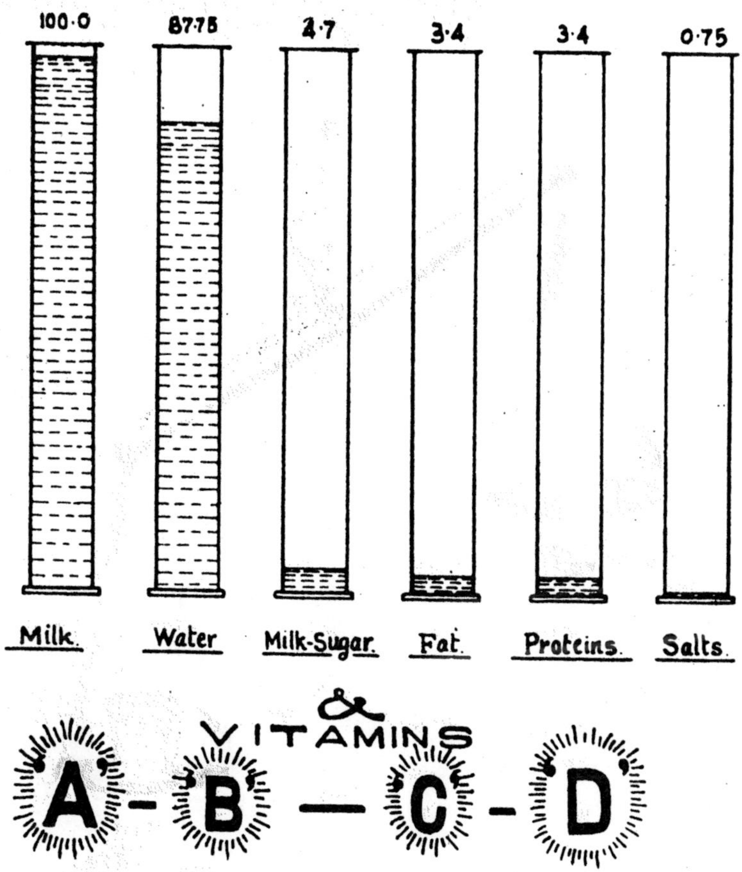

MILK

More Nourishing Than Eggs.

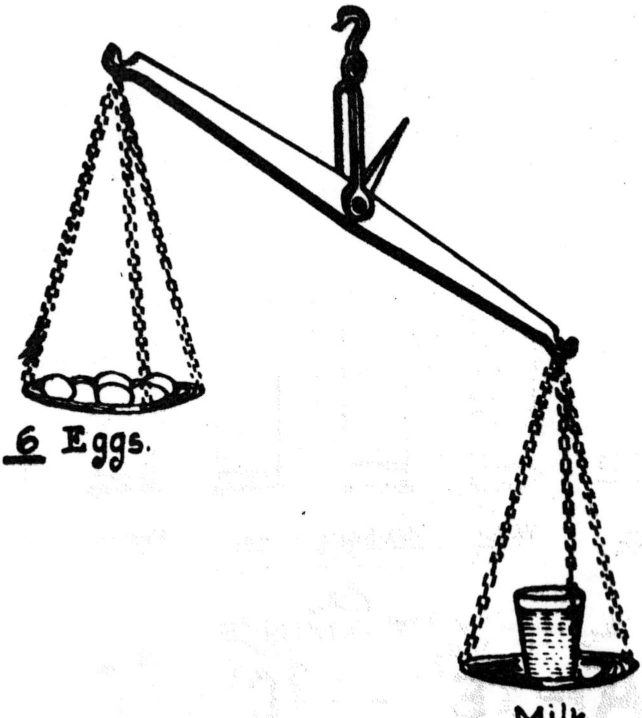

Chemical composition of Milk.

Mother's Milk—

Average Composition.	2·1 % Proteins.
	3·4 % Fat.
	5·1 % Sugar.
	0·25 % Salts.

Assimilation of Mother's milk takes place to the extent of

99 % Proteins.
97 % Fat.
90 % Salts.

Ass's milk is much used in France and in Germany. In Dresden there are regular establishments for its sale, but as the yield of milk is poor, it is very expensive.

Sheep's milk is recommended for children because it is supposed that the sheep rarely suffers from Tuberculosis.

From the above table, it is clear that the constituents vary from one animal to another, depending upon the nature of the animal and the requirements for the growth of its young ones. Again, even in the case of the same animal, it has been found that the constituents vary in their percentage contents, according to the requirements of its young ones, at different periods of its growth. It will be also seen from the above table, that in the case of hardy animals which have to do a lot of rough work, as is the case with the mare and the ass, the percentage of fat is always low.

The selection of a particular milk is to be made dependent upon whether the adult in question requires more percentage of fat or of other constituents. From this point of view, it is desirable to study the problem with special reference to the individual constituents.

Specific Gravity :—This property is the net result of the presence of the various constituents in their percentage proportions. But this is never a very reliable and constant property. The various milks can be generally classified from this point of view as follows :—

A comparison of Specific Gravities of various milks according to Godbole-Sadgopal.

Elephant	.. ..	1·2-2·75	Mare	..	1·03-1·042
Sow	.. ..	1·403	Goat	.. ..	1·03-1·033
Buffalo	.. ..	1·038-1·042	Cow	.. ..	1·026-1·035
Sheep	.. ..	1·035-1·042	Woman	.. ..	1·03-1·034
Bitch	.. ..	1·038	Ass	... ...	1·023-1·035

It will be interesting to note that the specific gravity of the milk rises on skimming off the lighter fat portion and, therefore, an extra addition of water is possible as adulteration, the resulting mixture giving the correct specific gravity figures within the range of those of pure milk.

Colour of the milk.

It is commonly believed that the colour of cow's milk is yellow and that of buffalo is what is known as "milky-white". The accuracy of this statement is very much questionable. As is well-known, milk is an emulsion of very finely divided particles whose size is also an important factor in determining the particular colour of the milk as studied from the point of view of colloidal chemistry. Again, some of the colouring matter is also derived from green grasses etc. on which the animals are fed. It is this colouring matter from the grass etc. which gives tinge to the milk and through the milk to the butter also. But this is only accidental. It has nothing to do with any physiological reactions involved in the building of milk in the system of the animal. If instead of straw or grass, the cow is fed on oilseed cakes, as is not very uncommon, the emulsion is white and not coloured. The yellow colour associated with cow's milk and cow's butter or ghee has almost become a superstition and as a consequence even the dairies use a yellow colouring matter in the manufacture of butter. The mere presence of a yellowish colour should not, therefore, mislead anybody in drawing the inference that necessarily cow's milk or butter is involved.

According to Neumann, winter butter is generally very nearly white because straw is added to the cake food given to the animal. If the animal is fed on carrots and other grasses, a yellowish colour, as also a peculiar aroma, is imparted to the butter which is then known as May-butter, Summer-butter or Grass-butter.

Fat-content.

The nature of the different fats obtainable from various milks differs in every case according to the nature of the animal, climatic conditions, feeding and idiosyncrasy of the animal etc. The classification of the fat-content is given in the following table.

A comparison of the fat-contents of various milks according to Godbole-Sadgopal.

Animal	Fat-content	Animal	Fat-content
Elephant	12·5-15·6	Bitch	4.2
Sow	9·78	Goat	3·2-3·95
Buffalo	6·5-8·75	Mother	3·0-3·85
Sheep	5·2-8·6	Ass	1·3-1·5
		Mare	0.6-1.8

The milks of American and European cows contain a lower percentage of fat than that of the Indian cows and that is why it is stated that, in the military and Government dairies in India, a portion of the fat from the milk is skimmed off and the cream removed with a view to bringing down the percentage content to between 3 to 4 and then the milk is sold as "Full Natural Milk"!! This also suggests a method by which the milk of the cow and the buffalo can be converted into a milk akin to mother's milk by simply treating the milk in an Alpha-Lavel-Separator and skimming off a portion of the cream so as to bring the fat content of the milk under consideration to that of mother's milk. The removal of the extra fat will correspondingly raise the content of the proteins and salts in the milk both of which are more desirable for the growth of the young ones. A subsequent addition of milk-sugar is then made to make it equal to mother's milk.

Regarding the percentage of butter in different milks, Kautilya (300 B. C.) in his *Arthashastra*, (2-29), mentions the following. "One *Drona* (equal to 6 quarts) of cow's milk will, when churned, yield one *Prastha* (equal to 3/8 quart) of butter ; the same quantity of a buffalo's milk will yield 1/5 *Prastha* more ; and the same quantity of milk of goats and sheep will produce 1/2 *Prastha* more."

Milk-sugar.—This sugar is easily digestible and on fermentation gives rise to acids like lactic, butyric etc. A small quantity of ethyl alcohol is also produced during the fermentation of this sugar as in the formation of Indian curds or *dahi* from milk. When milk gets sour, acids like lactic and butyric are produced. The digestive properties of butter-milk are mostly dependent upon the lactic acid formed during the fermentative process and hence its importance in the Indian diet can be understood. The following table will show the order in which the various milks can be placed from the point of view of their percentage content of milk sugar :—

A comparison of the milk-sugar content of various milks according to Godbole-Sadgopal.

Elephant	.. ...	7·2-10·3	Buffalo	.. ..	5·5·4
Mare	... ...	6·8·5	Sheep	.. ..	4·5·5
Ass	.. ...	6·3-6·8	Cow	.. ..	4·5-5·2
Sow	... ...	6·5	Goat	... ..	4·5·3
Mother	... ..	5·8-6·5	Bitch	.. ..	3·8

Cane-Sugar.

"An excessive proportion of sugar in the diet is especially bad for children, whose instinct to eat it seems altogether out of proportion to its nutritional qualities, since it occupies the place which should be filled by protective food. It also spoils the appetite for the latter, which are even more necessary during the period of growth than in adult life."

Proteins.

In general, it is known that the proteins of animal origin are more easily digestible than the same from vegetable sources. For the same reason, the proteins present in Indian pulses are not so easily digestible as the nitrogenous matter from the milk casein or the yolk of eggs. Again the shorter the time taken by young ones to double their weights, the more the albuminous matter required for their growth. It is known that the greatest percentage of nitrogenous matter is to be found in the case of hare's milk—about 14.5%. In the case of milk, we have got three types of proteins—(*a*) Casein, (*b*) Lacto-albumin and (*c*) Lacto-globunin. The latter two are more easily digestible by the human system than casein. Casein again differs in its nature according to the sources from which it is obtained. For example, it has been found out that the casein in the case of mother's milk is less easily precipitated than the other caseins as in the case of elephant and buffalo milks. Again the casein precipitated from mother's milk is much lighter than in the case of buffalo's or elephant's milk showing thereby that mother's milk contains casein which is more finely divided and hence, also, more easily digestible. In the following table, the various milks are arranged according to their casein-content :—

A comparison of the casein-content of various milks according to Godbole-Sadgopal.

Origin :	Elephant.	Sow.	Buffalo.	Sheep.	Goat.	Bitch.	Cow.	Mare.	Ass.	Mother
% Casein	7.2-8.9	6.1	4.5-5.2	4-5.2	3.2-3.5	4.2	3-3.4	1.5-1.7	1.2-1.3	0.4-0.9
% Albumin + Globulin	3.1-4.5	2.5	0.8-0·95	1.8-2.3	0.4-0.7	0.5	0.4-0.6	0.6-0.85	0.4-0.7	0.6-0.75

It is very interesting to note that the percentage of milk-sugar is very large in the case of mother's milk although there is a deficiency in the percentage of albuminoids. It is therefore that children always like to take mother's milk as it is sweeter than the other milks which they are unwilling to take without some extra sugar. These albuminoids play a very important role in the building up of the tissues of the human body for which very reason, some of the popular medicines for the purpose (as for example, Sanatogen) are rich in casein content.

Ash

A study of the ash obtained by incinerating the solids and thus removing the organic matter is very interesting. Salts of the metals like calcium, magnesium, potassium, sodium, iron, containing chlorine, phosphorus and sulphur in a combined form are found to occur in various milks according to the nature of the animal and the food on which it is fed. Sulphur has been found only in cases of milks of the goat and the cow. Again iron has been found only in the case of cow-milk while traces of iodine have been observed in both the milks from the cow and the goat. Physiologically speaking, these salts, though present in very minute quantities, play a very important part in resisting a number of diseases like T.B. and development of tonsillitis. In this connection, it should be remembered that due to the presence of these various salts of organic nature, the normal milk is amphoteric in its reaction, *i. e.* it behaves both as an acid and also as an alkali towards an indicator.

Zbinden has recently observed, during a thorough spectroscopic examination of milk, traces of the following metals : aluminium, chromium, copper, lead, manganese, tin, titanium, zinc and vanadium. In the opinion of this investigator, even traces of these metals play a very important part in imparting to milk the best nutritive and medicinal properties.

Mother's milk contains traces of silver also in addition to other metals as proved by the spectroscopic examination.

Mother's Milk vs. Artificial Feeding.

Of late, the importance of feeding children on mother's milk is being minimised. It is being considered fashionable by educated (!) mothers that their children should be fed artificially, either on cow's milk or artificial diet-milks. It has been ascertained by practical experiments conducted by the Infant Welfare Centre, Chicago, that *breast-feeding of infants* is of very great importance because it is cheaper, cleaner, healthier and safer than *artificial feeding.* In Chicago, during the years 1924-29, 20061 infants attending the centre were closely followed up for the first nine months of each infant life. Of these, 48·5 per cent. were wholly breast-fed, 43·0 per cent. partially breast-fed and 8·5 per cent. wholly artificially fed. All the infants—artificially fed and otherwise—were attended at intervals by the officials of the Centre. The mortality rates of these different groups of infants were as follows :

	Number of infants.	Total deaths.	Percentage of deaths of infants.
Wholly Breast-fed	9,749	15	0·15
Partially "	8,605	59	0·7
Artificially fed	1,707	144	8·4

It will be seen from the above that the rate of mortality among the artificially-fed infants is fifty-six times greater than that of those completely breast-fed. The difference in the death rate between these classes was largely due to deaths following respiratory infections and to a less degree, gastro-intestinal and other infections. Thus, whereas only four out of 9,749 of the breast-fed infants died of respiratory infections, eighty-two out of 1,707, artificially fed infants (*i.e., to say nearly 480 infants out of 9749* !) died from this cause.

The "food-peril" due to the infected milk of an artificial diet and the diminished degree of resistance to infection produced in infants fed on *artificial diet* as compared with those receiving mother's milk are the two main causes which are responsible for the increased frequency of respiratory gastro-intestinal troubles and consequent mortality in the case of artificially fed infants.

VIII

Vitamin Contents of Milk

Presence of vitamins in milk is responsible to a very great extent for making it an excellent health-tonic. According to the best authorities on the subject of vitamins practically all the vitamins and especially vitamins A, D and E are abundantly present in the milks of various animals, the latter being prominent in case of the milks of the cow and buffalo. In the case of mother's milk, the vitamin potency varies according to the nature of the feeding. It is therefore that wise mothers always take special care of their food during the period of their pregnancy. Medical workers in Europe have found out that vitamin-potency of mother's milk is practically nil, in cases where the diet of the mother consists only of a non-vegetarian menu. Even in the case of other animals it is found that vitamin-content is high in case of those animals which take the largest amount of greens, grasses and vegetables. A very strong argument in favour of milk is to be found in the fact observed by American physicians that sterility in both sexes is generally found in those people who seldom take milk. Again, the vitamin-potency of fresh milk of the cow and the buffalo (commonly known as धारोष्ण) and also of the fresh milk of the mother is the highest in comparison with the heated milks as the heat destroys many a vitamin. In India also children are given fresh milk (धारोष्ण) of goat by educated mothers and such children have been generally found to add to their growth very rapidly.

IX

FOOD VALUE OF MILK

Compared with Eggs, Meat etc.

One litre of Milk contains *units of Calories.*—

Equivalent to	600	gms.	of	Beef.
	400	gms.	of	Pork.
	750	gms.	of	Calf-flesh.
	8	pairs	of	,, Sausages.
	9	eggs	of	Hens
	2·6	kgs.	of	Cauliflowers.
	1·4	kgs.	of	Apples.
	2·0	kgs.	of	Green Beans.

In Germany, for a regular daily food of a grown-up man in calories and nourishing materials, the cost comes to *1·2 R. M.* in terms of *Milk*.

But in Herring (Fish)	1·9 R. M.
Sausage	3·5 ,,
Ham	3·9 ,,
Pork with bone	4·2 ,,
Eggs	4·4 ,,
Beef	5 0· ,,

Food value of Eggs:

Composition of eggs: Water 74% +Albuminoids 14·1% (: 7% assimilable) + Fats 10·9% (95% assimilable).

On keeping, eggs lose their water gradually and are liable to bacterial decomposition which is certainly harmful to the system. Fresh eggs sink in a 5·1% solution of common salt whereas the stale ones float on it. In the tropical climates particularly it is risky to use eggs as food materials because decomposition sets in very quickly by the bacteria travelling through the outer shell into the inner egg. In European climates where the external temperature is sufficiently low (when compared with the Indian tropical climatic conditions), special precautions are taken to store the eggs in cool places and the sanitary authorities stamp every egg with their seal of purity before they are sent to the market. The danger of using eggs cannot, therefore, be exaggerated in tropical countries like India, particularly in provinces like the Punjab, U. P. or Sind (Jacobabad maximum 130°F, where it is as hot as hell) and where there is neither state control nor sanitary supervision over the production and distribution of eggs. When compared with milk both in price and food value, eggs are, undoubtedly, dangerous in India.

The shell of the egg constitutes 1·1 % of the weight of the whole egg, the yolk 32 % and the white 57 %.

	Refuse	Water	Protein	Fat.	Ash.	Cal. per lb.
Whole egg	12·2%	65·5%	11·9%	9·3%	0·9%	635
Whole egg.—edible portion		73.7%	13·4%	10·5%	1·0%	720
White of egg		86·2%	12·3%	0·2%	0·6%	250
Yolk of egg		49·5%	15·7%	33·3%	1·1%	1705

" Eggs are liable to deterioration because the shell is porous and water evaporates through this covering. In its place, air rushes in and causes decomposition of the organic matter of the egg. Various gases are formed—principally sulphuretted hydrogen, a nasty gas generated by the action of putrefactive bacteria which enter the shell with the air. Eggs, eaten in this state, may cause gastric and intestinal disorders—particularly so in tropical climates "—where in summer the temperature reaches almost 110°-120°F.

1 average egg	yields	60	Calories.
1 ,, white of egg	,,	13	,,
1 ,, yolk of egg	,,	48	,,

9 to 10 eggs weigh 1 lb.

Plant foods compared with Animal foods.

Plant foods contain as a class the five main constituents, *viz.*, proteins, fats, carbo-hydrates. mineral matter and water (besides vitamins) like animal foods but they

BUTTER-MILK

ITS COMPARATIVE NUTRITIVE VALUE.

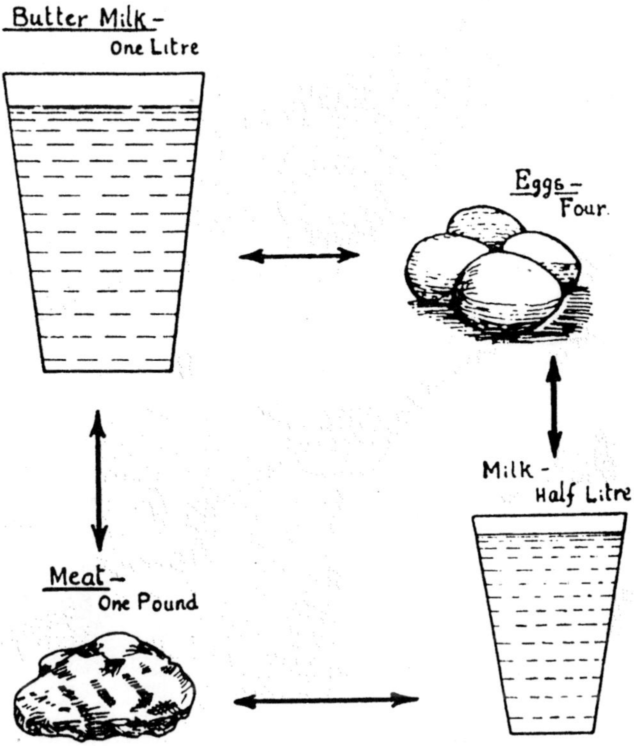

Equivalent to 4 eggs is one litre of butter milk.
Equivalent to 1 lbs. Meat " " " "
Equivalent to ½ Litre Milk " " " "

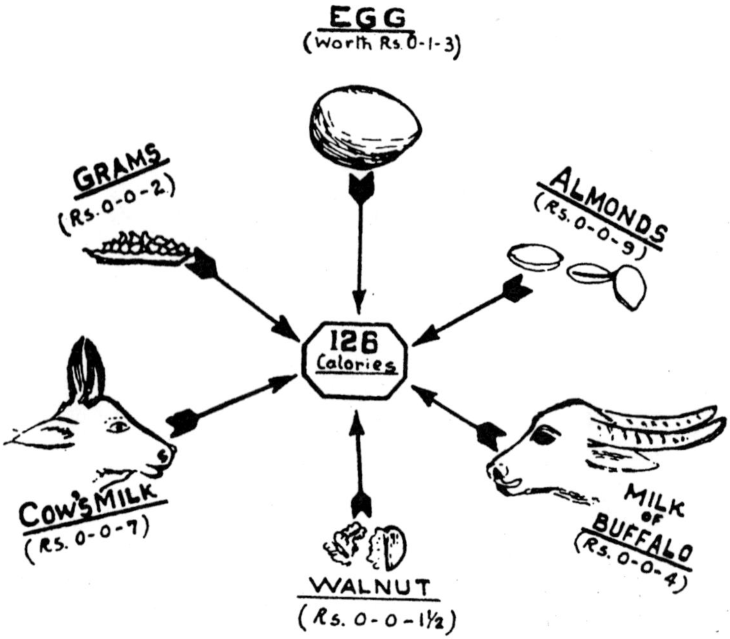

EFFICIENT SUBSTITUTES FOR EGG.

differ in the proportion of the constituents, the latter foods being, as a rule, richer in proteins and fats and poorer in carbo-hydrates than the former.

The following table makes this point clear.

	Prot.	Fat.	Carb.
Beef free from fat	23·2	2·5	
Lean beef, very fat.	16·1	23·1	
Wheat flour (entire)	13·8	1·9	71·9
Potatoes	1·8	0·1	14·7
Apples	1·6	2·2	66·1

Legumes, Beans, Nuts are rich in proteins, or fat or both but, unlike meat, they are rich in carbo-hydrates also.

	Pro. %	Fat %	Carb.%
Beans dried	22·5.	1·8	59·6
Peas	24·6	1·0	62·0
Almonds	21·0	54·9	17·3
Walnuts	16·6	63·4	16·1

Soya Beans.

In the natural order *Leguminocae*, there is a set of valuable cereals which are of great value to the vegetarian particularly and to the non-vegetarian also. The peas, different kinds of beans, soya beans etc., are rich in proteins, even richer than meat and flesh. There are still in India varieties of these beans, growing in different parts, which need investigation from the chemical point of view. The nature of the proteins and their assimilability from a physiological point of view also need a careful investigation. It is now established beyond doubt that soya beans are not only not foreign to India but two varieties of these are known to be indigenous to the Kumaon district, for several centuries. As a class, the vegetable proteins are known to be less assimilable than meat-protein. But it is worth investigating, whether they could not be rendered equally assimilable, provided they are boiled for some time, perhaps in the presence of an alkali like sodacarbonate, used more in the form of powders than in the form of broken or whole grains and allowed to soak in water for several hours before use and even allowed to germinate slightly. Soya bean used in this way is a rich food. It contains vegetable proteins to the extent of 33% and 17% of fats. In the form of soup or pastes these and other beans are highly useful. In the last European War, these beans were used in the form of fine powders, as substitutes for flesh preparations. These vegetable proteins are protected by cell-walls which if not properly broken up and disintegrated render the proteins less assimilable than otherwise. In the form of well-extracted soups and finely-powdered pastes, the assimilability can be highly increased and made equal to that of meat proteins.

Analysis of Soya Beans.

Moisture	5·3 %	Proteins (N × 6·25) =	37·0 %
Fats	16·8 %	Carbohydrates =	31·0 %
Cellulose (fibre) =	5·1 %	Ash =	4·8 %

X

INFLUENCE OF MILK ON THE GROWTH OF THE BODY AND LONGEVITY OF LIFE

Milk and Height-growth.

It is quite well-known that milk is not only a good nourishing material but is extremely valuable in increasing height, general cheerfulness and resistance to disease. Experiments on boys and girls have been tried in many places, but the following experiment conducted by Dr. H.C.C. Mann, an English doctor, as communicated to the British Medical Research Council, is very convincing. The experiment was tried on 500 boys and over a period of four years. The trials have shown that the additional supply of one pint of milk, in addition to the normally calculated efficient food supply, has shown an additional increase in weight of 1½ to 3 Kg. and an increase in height of about 4½ to 5½ cms. The children selected were orphans generally found in the city of London and fed in poor-houses. These orphans were previously subjected to a medical health examination and all doubtful cases were excluded. They were under the observation of the medical authorities most of the time. They were divided into several groups and one of these groups was fed on the routine nourishment—sufficient to meet the usual routine demands—fitting the age of the boys. This nourishment, if any, was even better than that supplied in the English poor-houses. Imperial doctors certified that the nourishment supplied was quite satisfactory. Of the other groups, one was given an additional diet of sugars, another an additional diet of albuminoids and still another an extra diet of margarine, yet another an extra diet containing butter and the last an extra diet of milk. Experiments proved that the butter and the milk groups showed the most remarkable progress. Any stray visitors could easily single out the children fed on the butter and milk group because of their markedly healthy physique. It was interesting to note that when all the other groups suffered from a very bad cold, the milk group remained quite normal and healthy and easily resisted the cold.

Dr. Hyslop Thompson, Medical Officer of Health for the Herts County Council, describes the uses of milk in illness, in a small pamphlet issued by the Agricultural Educational Sub-Commmittee of that county, in the following terms :—"In certain wasting diseases, in digestive disorders, in acute diseases, associated with high fever and the nervous conditions which result from the stress and strain of modern life, milk provides a form of nourishment pleasant to take, easily assimilated and of proved value and benefit to the patient". In 1926-27, large-scale tests were carried out in schools in several cities and towns in Scotland and Belfast under the auspices of a Committee appointed by the Scottish Board of Health to determine the nutritive value of milk for children. At each centre of experimentation, four groups of children were taken, each numbering from 40 to 50 according to the size of the classes. One group received "whole" milk. A second group received separated milk. A third group received biscuits of the same energy-yielding value as that of the separate milk and the fourth group which acted as "control" received no supplementary feeding. The tests began at the end of November and finished at the end of June. The five to six year old children received ¾ pint of milk per school day; the 8 to 9 year old, one pint; and the 13 to 14 year old, one and a quarter pint. These experiments showed that the children fed on milk had glossy hair and clearer complexions and held themselves more erect than those who went without milk. The general results were as follows :—

(1) The addition of "whole" milk to the diet of school-children during the 7 months' experimental period had been accompanied by a rate of growth as indicated by an increase in height and weight 20 % greater than that in children not receiving extra milk.

(2) This increase in rate of growth had been accompanied by an improvement in general condition of many of the children receiving milk.

(3) Even separated milk is of great value for promoting growth. Its nutritive value for children would appear to be under-estimated.

Long life and milk.

It is interesting to note the relation between long life and the habit of taking plenty of milk. It has been ascertained in Germany that out of a total population of sixty millions, only 100 inhabitants are aged 100 years or more. It has been proved from the data collected by the Pasteur Institute at Paris, that in Bulgaria, with a population of about five millions only (*i. e.*, about one-twelfth of the population of Germany) nearly 5000 have a life of one hundred years or more. The reason for this is undoubtedly the nature of the food the Bulgarians take. The Bulgarians take a very large amount of milk, cheese, joghurt (*Dahi*) in their daily food, whereas flesh diet is taken only on holidays.

XI

CURD (*DAHI*) AND BUTTER-MILK.

Joghurt, Kumys and Kefir.

In India, the art of fermenting milk into "*Dahi*" is centuries old. The underlying principle of converting milk-sugar into its fermentation products—such as lactic acid, alcohol, carbon-dioxide etc., is quite well-understood. Every house-wife starts with hot milk and adds to it a little of the previous day's *Dahi* or "Lasi" (Butter-milk) and allows it to ferment for 12 to 24 hours as the case may be. As the previous day's ferment is not a constant substance with one special type of bacteria, the resulting product *Dahi* naturally varies from day to day. It is here that a little scientific knowledge is necessary to prepare a uniformly standard product. Particularly in the extreme winter days—whether it is in the north or south or east or west of India—every household experiences the difficulty in preparing *Dahi* of a proper quality. Either the fermenation does not take place at all, or if it does, it gives a broken and curdy mass but very rarely a uniform gel with a standard taste. In fact it becomes at times too sour (a result of influence of high temperatures) or remains too sweet also. In this matter a very scientific study has been made of the subject—particularly in Bulgaria and Germany. Under the name of "Joghurt" (*Dahi*) in Germany, a number of Ph. D.s are carrying on a regular business of manufacturing *Dahi* from day to day under exactly accurate and scientific conditions so that no matter, whether it is winter (−30°C) or summer (+15°to20°C) absolutely one uniform product of one uniform taste and composition is available all the year round. Particularly in Bulgaria, all kinds of milk—sheep's, goat's, cow's or buffalo's (wherever available) are used for this purpose. The method adopted is as follows : A particular ferment, known commercially as "MAYA" is used for the purpose. The active bacteria concerned in this ferment are known as "Bacillus bulgaricus", also partly "Bacterium lactis acidi" and a Streptococcus, all collectively known as "MAYA". These bacteria are now being manufactured and sold commercially by special milk institutes in Germany. In a little pasteurised milk,the bacteria are allowed to culture for about twenty hours at a temperature of about 45°C and to this is added the bulk of the milk to be fermented, previously heated to the same temperature of 45°C (at times, even concentrated at this temperature). The ferment added should form about 2–5% of the total quantity of milk. It is the allowed to ferment at a temperature of 40°–48°C for about 5–6 hours by which time about 30–35% of acidity is produced. It is then allowed to cool or cooled quickly by ice-cold water and is ready for use. The Joghurt or *Dahi* so prepared is supposed to be excellent for disinfecting the alimentary canal by destroying the bacteria that are responsible for the intestinal troubles. The proverbial long life of the Bulgarians as a race is attributed to the daily use of *Dahi*. We, in India, could make ourselves equally long-lived with our hereditary knowledge of the subject of *Dahi* or butter-milk, provided we standardise our product with the help of modern knowledge.

Kumys (Milk-wine) is a fermented liquid obtained from the milk of mares and used very largely in South Russia and Inner Asia. The large amount of milk-sugar present in mare's milk (6-8·5%) is fermented to lactic acid, alcohol and carbondioxide and is supposed to be very healthy for patients suffering from lung diseases and anæmia.

Kefir is a drink very much used in Caucasus and is prepared from cow's milk. The products of fermentation are butyric acid, acetic acid, lactic acid, alcohol and carbon-dioxide. This is supposed to have an acidic, pleasant and refreshing taste. A similar drink is used in Egypt under the name of "LEBER".

Butter-milk

Butter-milk is a by-product of the milk industry obtained after the butter is removed. The art of making butter and butter-milk is known in India from times immemorial. Butter-milk is obtained either by the sweet method (direct churning) or the acid method (after fermentation to *dahi*). The butter-milk obtained by the sweet method takes a little bitter taste after standing. The average composition of butter-milk is as follows :

Water	.. ..	91·0%
Fat	.. ..	0·5%
Proteins	... ..	3·5%
Milk-sugar and lactic acid	...	4·0%
Ash	... ...	0·7%

The butter-milk should be used as far as possible as soon as it is freshly made and should not be unnecessarily diluted. This precaution should be taken particularly in tropical climates where the high temperature is likely to give rise to new types of fermentation-products not always healthy for the human system. Butter-milk, if taken when it is freshly made, is a very valuable and nourishing drink.

Butter-milk contains less fat than milk and also contains 3–4% of proteins completely digestible, whereas that of milk is not so fully digestible.

Butter-milk is the cheapest nourishing material as is clear from the following comparison :

Articles.	Price of one kilo-gramme.	Nourishing units for one Kg.	Cost of 100 nourishing units.	Nourish-ment for one Rupee.
		Calories.	as.	Calories.
Eggs ..	As.15	890	more than 1½	944
Beef ..	„ 8	1225	more than ½	2450
Pork ..	„ 9	1749	nearly ½	3498
Skimmed milk.	„ 1	214	nearly ½	3424
Butter milk.	„ 1	224	less than ½	3584

XII

BUTTER, ITS NUTRITIVE VALUE, COMPOSITION AND DIGESTIBILITY.

The nutritive value and the digestibility of butter-fat.

In India, one of the very important questions often discussed is how far tallow, lard, vegetable ghee or a similar hydrogenated oil can be considered suitable for nourishing the human body. One prevailing idea is that such fats are positively harmful to the human system, whereas the other view is that these may not be harmful but surely they are not so nourishing as butter-fat is. In expressing an opinion on this vexed subject, it would be best to put together all the information which could be gathered from various authoritative sources.

In drawing inferences from the physiological observations and experiences of the scientists on the subject in Europe and America, one important aspect has got to be borne in mind and that is, that both in Europe and in America, butter-fat as such does not play a very important part in daily diet. The competition in those countries is between pure butter on the one hand and synthetic m r-garine on the other. For the daily use on the breakfast table what is required is either butter or margarine, both of which on account of their semi-solid consistency have got the property of being spread on the bread. The rich and the poor alike have the same type of menu, the difference being one of quality. The poor cannot afford to pay for the expensive butter. Margarine, *i.e.*, an emulsion of oils and fats with water and a flavour of butter takes its place. In America, particularly, a very interesting and serious controversy is going on between the margarine and butter schools of thought. A certain amount of data is available, but, on the whole, it appears to be more of a propagandist type than of a purely scientific nature. The number of factories manufacturing margarine and the sums of capital locked up in their working are so large that the margarine school is naturally endeavouring to hold its own against the established popularity of the butter school. The researches of the margarine school regarding the physiological effects of margarine in the human system have got to be carefully interpreted, before they could be made applicable to the oils and fats as used in India in their dehydrated form. The popularity of natural butter as compared with that of margarine and the preference given to its use even in the largest margarine manufacturing countries, in the face of a regular and enormous propaganda made in favour of the latter, is in itself a strong argument in favour of butter.

The views of the margarine school can be briefly summarised as follows: "According to the best scientific information, a pound of one kind of fat yields exactly the same amount of heat and muscular energy as a pound of any other kind of fat, and one kind of fat is digested with practically the same ease and completeness as any other kind. On the basis of actual utility of a fat as a food, there is, therefore, no choice except in the matter of cleanliness, soundness, freedom from contamination, price, convenience in handling, suitability for certain cooking purposes and last but by no means least, palatability." It should be emphasised here that the above statement refers to margarine which is not identical with animal fats, vegetable ghee or other dehydrated fats. The inference, therefore, cannot necessarily be the same.

As against this, the German school of chemists and physiologists have given us more useful data in the form of their experiments on the dehydrated oils and fats. These researches interest us because they throw proper light on the immediate problem before us. According to Koenig, the oils and fats which are easily decomposable are also easily absorbed by the human system. He found that by taking butter, margarine, lard, cotton-seed oil, sesame oil etc., there was no appreciable difference in the velocity of saponification of these fats. This argument of drawing an inference regarding digestibility from the saponification velocities is open to one serious objection and that is that in estimating saponification values, half normal alcoholic solutions of caustic potash are used, which are undoubtedly very strong alkalies, and which cannot be compared with the small (though active) quantities of ferments in the human system. This analogy, therefore, does not hold good. It has also been proved that the fats with the higher melting points are less digestible than those with lower melting points. Taking even the free fatty acids by themselves (which are invariably present in all Indian oils and fats, since the process of chemical refining is hardly known), it is found that the higher fatty acids, like palmitic and stearic acids, pass off through the human system unused and unabsorbed, whereas free oleic acid is taken up and absorbed by the system and by its mere presence helps partly to absorb the otherwise indigestible acids like palmitic and stearic.

Accordingly, it is found that oils and fats, whose melting points lie below the normal temperature of the body, are absorbed and digested in the intestines almost up to 97% to 98%, whereas in the case of tri-stearin (which is, by the way, the major constituent of animal fats and hydrogenated oils like vegetable ghee) it was found that only 9% to 14% was used, that is to say, the remaining 81% to 86% of the tri-stearine only helped in loading the system without in any way benefiting it. The assimilability of the acid as glyceride is, in general, a function of its molecular weight, *i. e.*, the lower ones are more easily absorbed than the higher ones; although in the case of unsaturated acids like oleic acid, they are more easily absorbed than similar but saturated acids both having the same number of carbon atoms. This appears to be a highly scientific reasoning because this relation of a uniform change of chemical and physical properties with an increase in the molecular weight fits in with a long-established theory of chemical constitution. From the above-mentioned references, it is clear that looking from the point of view of the chemical composition of butter-fat, animal fats, vegetable ghee, coconut oil, sesame oil etc., so far as digestibility and absorbability by the system are concerned, these can be placed in the following order :

(1) Butter-fat, (2) coconut oil, (3) other oils rich in oleic glycerides such as sesame and safflower oils etc., and (4) animal fats and hydrogenated oils of vegetable origin (vegetable ghee) or of animal origin.

In order to show the superiority of the butter-fat over other fats commonly used, the following explanatory tables are inserted:

A comparison of the melting and solidifying points of various fats.

No.	Fats.	Melting Point°C.	Solidifying Point°C.	Observers.
1	Mutton tallow ..	44-55	32-45	Holde-Bleyberg.
2	Beef tallow	40-50	30-38	" "
3	Lard	28-46	22-32	" "
4	Butter-fat	28·5-43·5	15-28	Godbole-Sadgopal.
5	Coconut oil	20-28	14-25	Holde-Bleyberg.

Both the oils and fats, whether of animal origin or vegetable origin, are mixtures of mixed glycerides of fatty acids of low and high molecular weights. In these again, the glycerides of high molecular weights are of either saturated acids or of unsaturated acids. Whenever the oils and fats are introduced into the human system, their assimilability and digestibility are dependent upon the nature of the acids. The glycerides of saturated acids of high molecular weights are not digestible and therefore load the human system and pass off undigested and unassimilated. They are likely to do, therefore, some harm to the system and no good whatsoever. This is mainly due to the fact that the glycerides of saturated acids of high molecular weights have melting points higher than that of the body temperature itself. According to Lewkowitsch, the melting points of the saturated glycerides of high molecular weights vary between 56·5°-75°C. Therefore, substances like mutton and beef tallows are the least useful if not also the most undesirable for feeding the human system. The liquid glycerides of the unsaturated acids of high molecular weights, particularly those of oleic and linoleic acids, on the contrary, are not harmful as those of the corresponding stearic glycerides of about the same molecular weights. These are easily absorbed and are known to be beneficial. The glycerides of saturated acids of low molecular weights ranging from those of butyric to those of myristic acids are known to be easily assimilated by the system, because of their very low melting points and easy digestibility and assimilability. The remarks made above for the non-assimilability of the solid glycerides of the saturated acids of high molecular weights are modified to a certain extent when the glycerides of either unsaturated acids or acids with low molecular weights are mixed with them. The proportionate percentage of the glycerides of saturated acids of high molecular weights on the one side and that of the glycerides of either unsaturated acids or acids of low molecular weights on the other decide the comparative assimilability and digestibility of any given oil or fat. The presence of free fatty acids, in general, in any given oil or fat, either arising out of the natural causes or as products of rancidity is always harmful to the human system. That is why in Europe every oil or fat used for edible purposes is refined for removal of the free acids before it is sold in the market, or before it is used for the manufacture of margarine. The following table, it is hoped, will throw light on the influence of composition on the digestibility and assimilability of the fats taken above for comparison.

A comparison of non-assimilable constituents of various fats.

No.	Fats.	Percentage of		References.
		Stearic acid.	Palmitic acid.	
1	Mutton tallow ...	25-35·5%	24-27%	Holde-Bleyberg.
2	Beef tallow.. ..	24·5%	27-29%	,, ,,
3	Lard	7·8-15%	24·6-32·2%	,, ,,
4	Butterfat	10-12%	26-31%	Godbole-Sadgopal.
5	Coconut oil... ..	0·8-5·0%	4·3-7·5%	Holde-Bleyberg.

A comparison of assimilable constituents of various fats.

No.	Fats.	Percentage of Oleic acid.	Percentage of Linoleic acid.	References.
1	Mutton tallow ..	36-43%	2·7-4·3%	Holde-Bleyberg.
2	Beef tallow.. ..	43-44%	2·6%	" "
3	Lard	50·4%	10%	" "
4	Butter-fat	30-34·5%	4-5%	Godbole-Sadgopal.
5	Coconut oil.. ..	10-10·2%	1%	Holde-Bleyberg.

A comparison of easily assimilable constituents of various fats.

Serial No.	Fats.	Percentages of Butyric.	Caproic.	Caprylic.	Capric.	Lauric.	Myristic.	References.
1	Mutton tallow	..	..	..	..	..	2-4·6	Holde-Bleyberg.
2	Beef tallow ...	..	..	..	..	..	2-2·5	Holde-Bleyberg.
3	Lard ..	..	..	..	..	...	...	Holde-Bleyberg.
4	Butter-fat ...	4	2	0·9	2	4-4·5	9-10	Godbole-Sadgopal.
5	Coconut oil ..	...	0·2-2·0	6-9·5	4·5-10·7	45-51	16·5-20	Holde-Bleyberg.

A comparison of the non-assimilable, assimilable and easily-assimilable constituents of the various fats.

Serial No.	Fats.	Non-assimilables. Palmitic+Stearic glycerides.	Assimilables+Easily-Assimilables. Oleic+linoleic glycerides.		Lower fatty glycerides.
1	Beef tallow ..	52%	46%	+	2%
2	Mutton tallow ..	55%	42%	+	3%
3	Lard ..	40%	60%	+	..
4	Butter-fat ..	36-44%	34-40%	+	22-23·5%
5	Coconut oil ...	9%	11%	+	80%

It is very interesting to note that human fat which is shown to be a synthetic assimilative product built up out of the digested part of the oils and fats taken in the form of food is known to possess the following composition and has a melting point between 15°-22°C.

COMPOSITION OF HUMAN FAT.

Stearic acid	4%
Palmitic acid	16%
Oleic acid	80%

It seems probable, therefore, that it is the lower and unsaturated glycerides which are primarily responsible for building up human fat.

XIII

CONDENSED MILK AND ITS VARIETIES.

Condensed milk and milk powder.

Besides being used for the daily house-hold purposes, in the form of milk, *Dahi*, butter, butter-milk etc., milk forms the starting point of a number of useful industries, such as the manufacture of casein and milk-sugar. In those countries, where there is a large over-production of milk, the excess of the supply over demand has got to be utilised in a form in which it will remain stable and useful. It is, therefore, preserved in different forms such as condensed milk, milk powder etc. In this form, it is used in places where it is not easily available as for example, on board the ships, in travelling or on the battle-front. This is only possible when the production is cheap and the demand is limited. In big cities, there is always a very great demand for pure milk and so the question does not arise. In villages and in places which have no railway facilities and where the transport difficulties are great, milk has got to be preserved. This is done by removing the large excess of water in suitable apparatus and under vacuum, adding sugar as a preservative and converting the full milk into a powder form.

The idea of rendering milk stable by removing the excess of water first originated in the United States of America (Hersford, Gail Burden) from where it was taken up to Europe where the idea developed on very scientific lines. A number of new processes were invented and patented until to-day milk powder and condensed milk are available in the market in excellent quality. The procedure adopted is as follows : Milk is concentrated by a process of boiling in vacuum at a temperature of 50°-60°C by the addition of about 12 parts of cane-sugar to one hundred parts of milk to the consistency of a syrup by reducing the whole to about $\frac{1}{5}$ to $\frac{1}{4}$ of the original volume. After cooling, the milk is packed in air-tight bottles or tins. The milk serves the purpose of a protective colloid. The milk can be concentrated to one-third of its volume and packed, even without the addition of sugar, as is often done. In this case, the quality of milk has got to be good and it should be as little acidic as possible. It should be noted that after opening one of these tins, the sugared variety can be used for a longer time than the unsugared variety which, therefore, must be used up as soon as possible.

Percentage-composition of Condensed Milk

	Sugared.	Not sugared.
Water	26·0%	48·5%
Fat	11·5%	16·0%
Albuminoids ..	12·0%	18·0%
Milk-sugar	16·0%	15·0%
Cane sugar	32·0%	..
Ash and salts ..	2·5%	2·5%
Total ...	100%	100%

New machines and new methods have been devised to prepare milk powder in a convenient form. The method of Krause is one of the latest in this connection and is worked on a principle whereby the milk is dried and removed within the least possible time so that in properties and composition it changes very little. Milk powder, in course of time, gets rancid, like oils and fats depending upon the atmospheric conditions. The composition of milk, skimmed milk, cream and butter-milk and whey is as follows:

Powders.

Constituents.	Cream.	Milk (full.)	Skimmed-milk.	Butter-Milk.	Whey.
Water ..	5·0	4.0	9·0	9·0	2·0
Fat ..	41·0	26·0	1·0	4·0	2·0
Albuminoids	21·0	23·0	33·0	17·0	14·0
Milk-sugar..	28·0	41·0	50·0	67·0	74·0
Salts ..	5·0	6·0	7·0	3·0	8·0

Composition of Colostrum milk.

Colostrum milk is the milk which is given immediately after the calving period. The composition of this milk is remarkable in many ways because it differs very much from the composition of the normal milk of the same animal. The colour is generally yellowish, sometimes slightly reddish, the consistency is sticky and the taste is saltish. The specific gravity ranges from 1·04 to 1·08. The fat in the colostrum has a low specific gravity (0·865), a small Reichert-Meissl value, high refraction and high iodine value and a high content of lecithin and cholesterin. The average composition of this milk is as follows :

Water	71·7%
Fat..	3·0%
Protein	21·0%
Milk-sugar ..	2·5%
Salts	1·8%
	100·0%

Caution.

There is a very great danger in the use of tinned milk as well as tinned fruit, fish or meat in India. The danger of what is known as *Ptomaine poisoning due to the contact of the metal of the box and its action on the foodstuff* is common both to the European and Indian climates but that due to the action of the great heat of the tropical climates is particularly serious in India. It is a matter of regret that no separate data are available of the deaths due to food-poisoning in India. It is also well known that, in India, *many young babies fed entirely on tinned milks and milk-products are known to die due to poisoning which is a result of the decomposition of the exposed tinned foods.* Tinned fruit, fish and meat are also known to cause similar casualties in the case of adults. The average low temperature of the cold countries induces but a very slow rate of decomposition in the food-stuffs there, whereas in the tropical climates, the rate of fermentation is very great and, therefore, the decomposition of food sets in very rapidly. Another serious cause of such occurrences is the fact that, when the air-tight tins are opened, the contents are not immediately transferred to other glass containers but are left in the tins as they are. In the case of milk-foods, also, sold in bottles, knowingly or unknowingly, the covers remain loose or open and decomposition sets in. It is, therefore, highly necessary to sound a note of caution and warn those who are given to the use of tinned foods in India against this serious evil. It is a great pity, indeed, that in a country like India, where everything "natural" can be had in plenty, the use of tinned or "synthetic" food stuffs should be getting so popular.

XIV

Cheese (*Panir*) its composition, manufacture and food value.

When whole or separated milk is curdled either by the action of an acid or by the action of rennet (obtained by extraction from the lining of the stomach of the calf and sold in the form of liquid or powder or tablets), the casein gets precipitated. The whey is then separated, the solid casein with or without the fat is pressed and allowed to ripen with the addition of a little common salt. A slow process of fermentation sets in, converting the milk sugar into acids and the casein into further decomposition products of an assimilable nature. The cheese obtained from the skimmed milk is less nourishing than that obtained from whole milk. The process of ripening takes from a few weeks to a few months after which the cheese is said to be mature. Four types of cheese are available in the market, made of (*a*) whole milk to which cream is added, (*b*) whole milk only, (*c*) skimmed milk, and (*d*) made of margarine with or without the addition of butter. Depending upon its consistency and the amount of water contained, it is either known as hard or soft cheese.

During the process of ripening, very complex chemical changes are supposed to be taking place which are still being investigated. The protein is partly decomposed, the milk-sugar is transformed and the fat undergoes partial splitting. A characteristic aroma is also developed depending upon time, temperature and hygrometric conditions. The taste for this has got to be cultivated, like the taste for beer or wines although to many, this aroma is positively repulsive. The manufacture of cheese is a regular science by itself, the whole process being under the regular control of a number of factors. It is sufficiently profitable to be a regular industry in India in view of the increase of imports of this article into India if a surplus of milk is available.

Cheese is a valuable food material, rich in protein, and is prized as such all over Europe and America. In a concentrated form, it makes a good food and in combination with bread, butter and beer is considered a delicacy in many countries. There is a very serious handicap in the manufacture or storing of cheese in tropical climates where the high temperatures affect the keeping qualities of this article. Whereas in cold climates, it can be safely stored for a length of time, it is difficult to keep it even for a few weeks either in the summer or the monsoons in tropical climates.

It is interesting to find the following reference to cheese and coagulated milk (precipitated by an excess of acid in the milk) in Kautilya's (300 B. C.) *Arthashastra*, 2-29. "They may also make use of coagulated milk or cheese to render their (cattle) oil-cakes relishing."

कीलाटो घाणपिण्याक क्लेदार्थः ।

Composition of Different Types of German Cheese according to Teichert.

Serial number.	Name of the cheese.	Variety.	% of water.	% of Fat.	% of Fat-free Dry Solids.	% Fat in total Dry Solids.	Ratio of Fat to Fat-free Dry Solids.
1	Emmentaler ..	Cream-cheese ..	34·0	28·5	37·5	43·2	1 : 1·32
2	Edamer	Cream-cheese ..	45·1	23·4	31·5	42·6	1 : 1·34
3	Tilsit	Cream-cheese ..	48·4	20·9	30·8	40·4	1 : 1·47
4	Camembert ..	Whole-milk	49·7	30·4	19·9	60·4	1 : 0·66
5	Muenster	Whole-milk	61·9	18·6	19·5	48·0	1 : 1·05
6	Romadur	Whole-milk	48·1	24·1	27·8	46·3	1 : 1·16
7	Limburger	Half-fat	55·2	12·7	32·1	28·4	1 : 2·53
8	Backstein	One-fourth Fat ..	59·8	7·4	32·8	18·4	1 : 4·44

XV

SKIMMED MILK, ITS COMPOSITION AND PROPERTIES.

Composition of Skimmed milk.

Skimmed milk is obtained by removing the cream either by the hand-method as is practised in villages or by the use of Alpha-separators as is done in dairies. The milk that remains behind is known as skimmed milk. The specific gravity of this milk is always greater than that of pure milk. In India, it is largely used for the preparation of *Dahi* and in this form, in big cities it fetches a good price and is a good food-material. The average composition of skimmed milk is as follows :

Constituents.	Hand-method. (Churning.)	Centrifuge method. (Alpha-separator)
Water	89·9%	90·4%
Fat	0·8 %	0·2 %
Proteins	4·0%	4·0%
Milk-sugar	4·5%	4·7%
Salts	0·8%	0·7%
Total ...	100·0%	100·0%

The specific gravity of this milk ranges from 1·0320 to 1·0365 at 15° the average value being 1·0345.

Composition of milk products.

.....	Total-solid matter.	Fat.	Proteins.	Milk-sugar.	Salt (Ash.)
Condensed milk (without sugar.)	26·5 %	7·8 %	7·0 %	10·0 %	1·5 %
Condensed milk with cane-sugar.	74·0 „	8·7 „	8·5 „	13·3 „ +41·0 „	1·5 „
Milk powder (Krause.)	97·0 „	26·0 „	28·2 „	35·5 „	7·0 „
Skimmed milk powder (Krause.)	95·0 „	1·0 „	37·0 „	48·5 „	7·7 „
Butter-milk ..	8·8 „	0·5 „	3·0 „	4·6 „ (Milk-sugar + lactic acid.)	0·75 „
Butter-milk (Dairy) Cold-skimming.	6·2 „	0·3 „	0·8 „	4·4 „	0·7 „

Vegetable milk.

During the European War, when there was a shortage of pure natural milk, a synthetic vegetable milk was prepared and introduced into the market under the name of Lahmann's Vegetable Milk. This was prepared from beans like Soya Beans and almonds and had a fairly thick consistency, brown colour and a pleasant aroma. The composition of this milk was approximately as follows :—

7% Vegetable proteins,

25 % Fat,

and 42 % Sugar.

One tea-spoonful of this vegetable milk added to a pound of ordinary milk and a pound of water makes a nourishing mixture poor in casein but rich in fat and sugar.

In China and Japan, a vegetable milk made out of Soya beans is being used by the local inhabitants for several years as a valuable substitute for natural milk. It is not difficult to prepare an artificial emulsion like natural milk because Soya beans contain all the ingredients although not in the same proportion as natural milk.

XVI

MILK, ITS PASTEURISING AND STERILISING.

Preserving and Pasteurising of milk.

Milk for household uses is kept and preserved for some hours before it is used. Wherever possible, it should be taken soon after it is milked, but this is not always possible. In dairies, it has got to be pasteurised before large quantities of milk are distributed over the city area. There are different types of bacteria that thrive in milk and they prosper in this healthy medium up to certain temperatures. Not every one of them is dangerous to life. The pasteurising, therefore, is adjusted according to the length of the period for which the milk is to be kept. A temperature of 10°C or below is good enough to keep the milk stable and healthy for some hours; but obviously this is not possible in the tropical countries. In the cold European climates, such a temperature is very common over a major part of the year. In big dairies, therefore, pasteurising is resorted to before the milk is distributed to the customers. Ordinarily, by pasteurising is meant the heating of milk in steam-jacketed vessels at a temperature of 60°-70° C for a period of 20-30 minutes. In the household, this is done by boiling the milk over a slow fire almost up to the boiling point. This pasteurising is enough to kill all the bacteria that are ordinarily harmful to the human system. The lower the temperature to which the milk is heated, the less the chemical changes that take place in the milk and the less the difference in its taste etc. The pasteurised milk is rendered stable under European conditions for a period of about 48 hours. The milk is cooled to normal temperature after pasteurising and is ready for distribution. A development of this idea has given rise to a new method known as "Biorising" where the milk under a pressure of 4 atmospheres is suddenly sprayed with the help of machinery (Biorisator) into a vessel heated to a temperature of 75°C. Here the contact of the finely divided sprayed milk with the hot surface is of a very short duration and yet the efficiency is much greater than that of pasteurising at 60° for half an hour.

Sterilising is another important process for preserving milk for almost an indefinite length of time. This is done either by heating the milk for 100° C for a length of time and removing all air and then bottling it or by heating the milk at a temperature of 120°-125°C in autoclaves whereby practically all types of bacteria are destroyed and the milk is safe for storing in bottles from which all air is evacuated. Such a milk, although it gets coloured brownish, is good for children in places where it is not available and also for being supplied to armies where the health of the fighting units is of extreme importance. This type of milk is known as totally sterilised milk.

A recent method of preserving milk is that of Budde who sterilises by the addition of hydrogen-peroxide (Perhydrol) to the extent of 0.35 gms. for a litre of milk at a temperature of 52°C. This addition does not bring about any change in the taste of the milk and it has been extensively tried with very good results in the last European war.

A new type of stabilised milk is what is known as "Homogeneous" milk. In this process, milk subjected to a high pressure is forced through fine nozzles and is dashed against walls of agate so that the fat particles of ordinary milk which vary in size over a wide range are broken up into smaller particles within a very narrow range. This sub-division of the fat globules increases the stability of the milk and retards the tendency of cream floating up. Incidentally, it makes both the milk and the cream look whiter than before. This fineness of the fat globules would naturally increase the capacity for assimilation due to the finer colloid particles.

XVII

VEGETARIANISM *vs.* NON-VEGETARIANISM.

One of the very important problems that taxes the brain of the average Indian is the problem of vegetarianism *vs.* non-vegetarianism. In regard to this question,there are a number of schools of dietetics preaching different combinations of edibles. (1) Firstly, there is the strict vegetarianism in India which would not include either eggs or fish or meat in the diet. It may be noted here that the staunch Buddhist in Japan would not include even milk in his diet because of its animal origin. (2) Secondly, there is the diluted European vegetarianism which does not object to eggs but rejects fish or meat. Bernard Shaw would be a classical example of this. (3) Next comes the group of honest non-vegetarians who practise different shades of non-vegitariansim which are only modified by religious belief, by individual taste or by convenience. The Hindu who would not touch beef, the Muslim who would not touch pork, the Chinaman who would not mind cockroaches and frogs, and all those for whom human flesh alone is the barrier, come under this category. Of course, the cannibal is in a class by himself. In the rational examination of the efficiency of this food-question, three important considerations are involved; firstly, the humanitarian aspect to which the Indian contributes in his own subtle manner; secondly, the scientific aspect of the question involving an investigation of the food values; and, thirdly, the aspect involving a comparison of the efficiency of food materials in terms of money. So far as the third aspect is concerned, *i. e.*, the question of efficiency with reference to cost, from the data available there seems to be very little doubt that the cost of an average non-vegetarian meal as available in Europe is more expensive than a similar meal under Indian conditions. Further, it has been the experience of this writer that during his travels in Central Europe, excepting certain big cities, it was difficult to get the vegetable dishes (except potatoes) without paying for both the meat dish and the vegetable dishes. There is no doubt that whether it be from the view point of food value or from the view point of cost involved, the vegetarian menu in India with the same amount of food-efficiency is in every way cheaper and healthier than a similar European menu.

India is a wonderful land. If there is one country in the world in which a part of the human race lives on purely vegetarian food, that is India. But even here there are various schools; for example, the first class is composed of those that will take pure vegetarian food (eggs and fish etc. excluded), then there is a second class of people that will take the vegetarian food, eggs and fish but will cry "Halt" at meat. The third class will go in for vegetarian food, eggs, fish and meat, including pork but no beef. Again there is the fourth class that welcomes all vegetarian food, eggs, fish and all other kinds of meat including beef but revolts at the sight of pork. And the last which may be called the Anglo-Indian type will include all vegetarian diet, eggs, fish and all kinds of meat. Some of these classes are convinced that their dietary is based on religious belief and it is out of question to try to dissuade them from their religious convictions. It is extraordinary that great religions are all based upon such convictions which do not permit of a logical reasoning or discussion. One common feature of all these schools of thought is that they all include milk and butter—both of animal origin—under the vegetarian diet, obviously because no killing (*himsa*) even in the embryo stage is involved in the production of these two valuable articles of food.

Before going into the idealistic humanitarian viewpoint of the strict vegetarian in India, a rational investigation of the food problem including the scientific, hygienic and the financial aspects would undoubtedly be very valuable. In this connection, it should be remembered that the non-vegetarian food, whether it be eggs, fish or flesh, is open to very serious hygienic objection. Even if the two diets are taken on a par, there is no doubt that the non-vegetarian diet is more dangerous than vegetarian food under similar conditions. *In a country like India, where high temperatures are common and where there is hardly any efficient sanitary supervision or legislation in the production and distribution of eggs, fish and flesh and where*

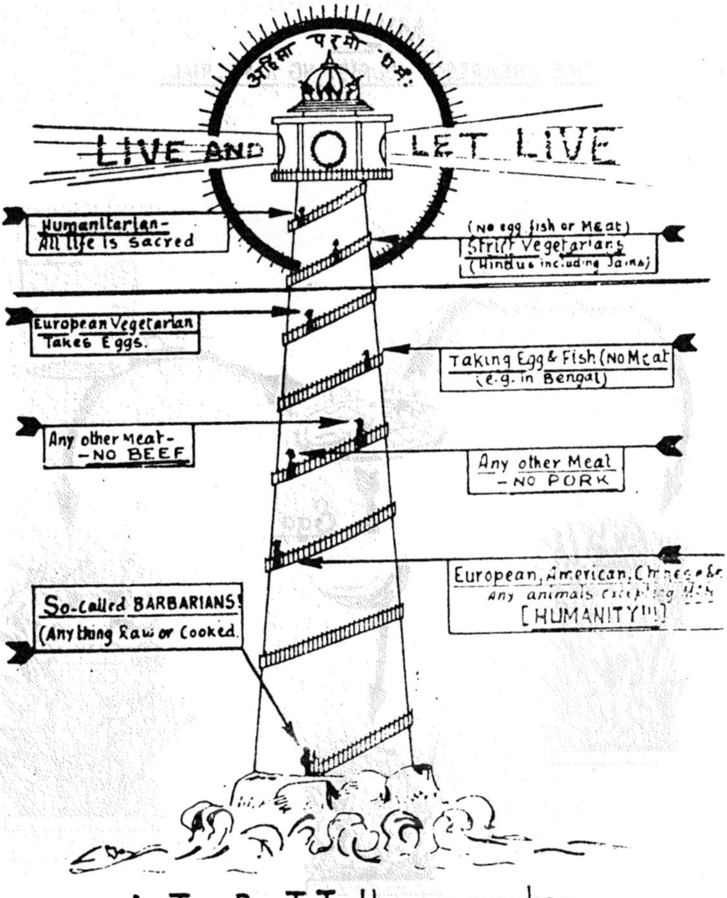

AS THEY RISE TO THE HUMANITARIAN IDEAL

MILK

THE CHEAPEST NOURISHING MATERIAL.

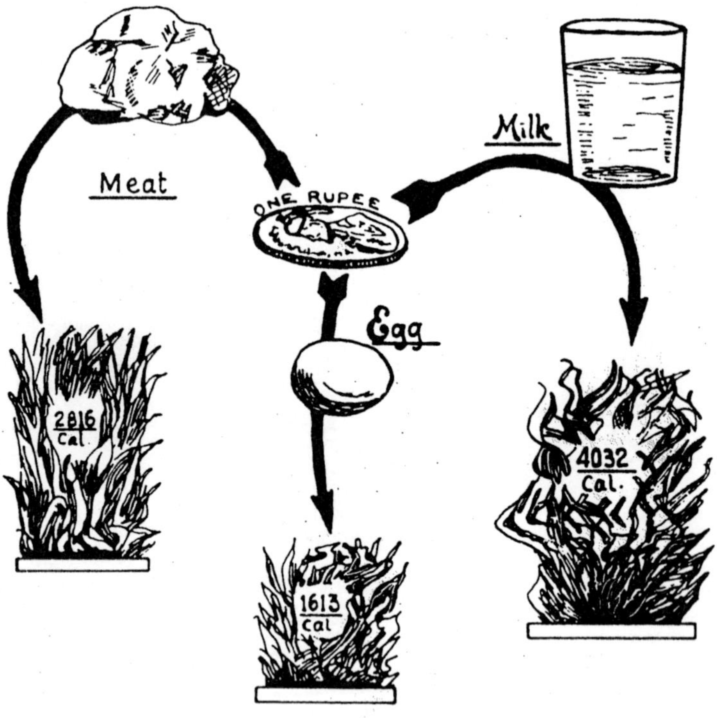

One Rupee worth of eggs yields 1613 calories.
One Rupee worth of Meat yields 2816 calories.
One Rupee worth of Milk yields 4032 calories.

anything like cold-storage of foods (as is the case in Europe and elsewhere) is a rare luxury, the dangers accompanying non-vegetarian food cannot be exaggerated. Besides this, it is to be noted that the vegetarian diet is not only much less dangerous but almost always more easily available and at correspondingly much cheaper prices, when the total food-value is taken into account. It has been shown in the West that meat is a carrier of dangerous diseases. Eggs and fish not properly stored are open to serious objection, the only point in their favour being that, *if* properly preserved, their assimilable nitrogen is of very great importance to the human system. The vegetarian in India, however, has in his favour an argument not less efficient when an alternative food supply giving a similar nitrogen efficiency is found primarily in milk and also in the different *dals*, beans and peas etc, (soya beans included). It is true that nitrogenous matter present in the vegetable world is embedded in a cellulose wall which has got to be broken up before the nitrogen is made fully assimilable, whereas with very little exertion, the nitrogen from the corresponding non-vegetarian food is more easily assimilable. It can be proved easily that in the vegetable world, with a little extra precaution of grinding and at times careful boiling, the nitrogen-efficiency can be fully utilised and can be as good as that of the non-vegetarian food. One of the very great experiments tried by Germany during the Great War, when there was a shortage of animal food, has proved beyond doubt that vegetable nitrogen sustained the German army as efficiently as the non-vegetarian nitrogen. It is recorded that large German armies were fed during the war on vegetable food consisting of peas and beans etc, very well powdered and mixed with fats and yet the efficiency of the armies was maintained at its top-level. Special attention, therefore, should be devoted to the composition of vegetable foods rich in nitrogen as adequate and efficient substitutes for animal foods. If cost and hygienic considerations are taken into account, the vegetarian dietary for its nitrogen value is both rich in variety and cheap in cost—both considerations of extreme importance in tropical climates.

Cost and food-efficiency apart, the argument of the staunch vegetarian should be appreciated from another lofty angle altogether. Man does not live for bread alone.

As a representative of the highest evolution of rational beings, he has to live for some ideals. The loftiest of these ideals, it cannot be denied is "To live and let live". This ideal is embodied in the great Samskrit motto अहिंसा परमो धर्मः "Not to injure the innocent is the highest form of religion". It is for this reason that the staunch vegetarian refuses to include in his menu anything like eggs, fish or meat. The typical Jain, who puts a cloth round his nose to breathe filtered air free from the atmospheric bacteria, is the representative of the extremely logical though unconvincing limit of this ideal. But the strict vegetarian stands on humanitarian ideals when he extends the conception of humanity from man to all creatures having life and feelings—the cow, the sheep, the pig and so on. The European conception of humanity, strangely enough, stops at human beings and does not include the life of animals like the cow, sheep or the pig (or the egg with life in embryo) which, it thinks, are outside the scope of humanity. The humanitarian in India looks upon the life of a cow that has yielded milk all its life-time with the same feeling with which he looks at the life of one of his own relatives and rightly too. This ideal is undoubtedly an ideal worth living for. If in addition to this, the question of food-efficiency, cost and hygiene all help in preserving the ideal, there is no reason why this view should not be considered as the loftiest one to be fought for.

According to eminent scientists, man would be placed under the class frugivorous animals both on the anatomical and the physiological basis. The parts and structure of the organs which have been naturally meant for carnivorous animals are not to be found in the case of human beings. We give below the views of some eminent authorities.

1. Sir Dr. R. Lancaster, K.C.B., F.R.S. wrote as follows in the Daily Telegraph (December, 1909) :

"It is very generally asserted by those who advocate purely vegetable diet that man's teeth are of the shape and pattern which we find in the fruit-eating or in the root-eating animals allied to him. This is true. It is quite clear that man's cheek teeth do not enable him to cut lumps of meat and bones from raw carcases and swallow them whole.

They are broad, square-faced teeth with four or fewer low rounded tubercles to crush soft food as are those of monkeys. And there can be no doubt that man fed originally, like monkeys, on easily crushed fruits, nuts and roots".

2. According to Dr. Alexander Haig, human metabolism (the process by which food is converted into living tissue) differs widely from that of the carnivora. The carnivora are provided with the means to dispose of such poisonous products as are contained in and are produced by the fermentation of animal flesh, while the human system is not so provided. In the human body the poisons are not held in solution but tend to form deposits and consequently are the cause of diseases of the arthritic group, conspicuously Rheumatism.

3. In the general opinion of doctors, *Rheumatism, Cancer and Consumption are chiefly produced by meat diet and vegetarian diet has been considered the first step in nursing the man attacked by them.* Influenza, which is believed to be more deleterious in cold countries is found in an aggravated form especially among meat-eating persons while those that are free from meat-diet are not so severely or rather not at all attacked by it. The effect of meat-eating is that it induces a constant stimulation rendering the subject of it more liable to colds or chills and the consequent depression and inability to recover from its effect.

4. According to Dr. F. W. Pavy, "A vegetable and fruit diet quickly cures all the evils of rheumatism and gout".

5. According to Dr. Gibson Ward, "Acid blood is the origin and sustaining cause of all rheumatic affections. Go to the shambles and apply a bit of the test paper to the juice of the flesh there, and you find it gives an acid-reaction. People who are content with the herb-bearing seeds and the tree-producing fruits for their diet cannot have rheumatic pains or fevers or their joints cemented together in rheumatic arthritis. This acid blood is fruitful of other diseases such as leprosy and scurvy etc."

6. Dr. Alexander Haig, M.A., M.D., F.R.C.P., states in his great work on "Uric Acid as a Factor in the Causation of Diseases", that the cause of all sorts of headaches is uric

acid and that meat produces it by introducing into the body and blood uric acid plus substances of the Xanthin group.

"Meat has a special value as a source of protein and iron. Its popularity is chiefly due to its appetising and stimulating properties ; but, *from the nutritional point of view, the protein of ordinary muscle meat is inferior to that of eggs or milk.*"

7. The following extracts from "FOOD" by Robert McCarrison (Director of Nutritional Research, Pasteur Institute, Coonoor, S. I.), would be found to be very interesting and instructive :

"If the food contains enough milk and milk-products then it is not necessary to eat flesh-meat at all. "Enough" milk is not less than one pint a day and a quart, if possible.

"Eggs are rich in fats, calcium, phosphorus and iron which are present principally in the yolk. *Though not so good as milk, they are useful substitutes for it when milk is scarce.*

"One ounce of any of the *dals* contains as much protein as one ounce of meat, nearly twice as much as one ounce of egg and seven times as much as one ounce of whole milk.

"But if we combine *dal* and rice with milk or meat or eggs or fish then the suitable proteins contained in these animal foods make it possible for our bodies to use to the best advantage the less suitable proteins contained in the *dal* and rice. *Dals* are eaten in various ways, and in the way they are cooked makes a great difference to the good we get out of them. The best way to use them is to grind them into meal and to make them into *chapatis* with *atta* or barley or other suitable cereal grain.

"The best *dals* are *Arhar* and *Mung*. Whole *dal* is much better than washed or split *dal*.

(It seems the writer has little information of the *dal* of *Urad* or *Mash* which is in no way inferior in its nutritive value to meat. Further, *Urad* is an excellent binding material for giving strength to the internal system and of the body).

"An ounce of nuts contains more proteins than an ounce of egg and five or six times as much fat. Nuts are rich sources of Vitamin B but they contain very little Vitamin A and no Vitamin C. They should always be eaten with and not at the end of a meal and should be well chewed".

Digestibility.

Vegetable foods (whether they be in the form of legumes or cereals) are as a rule less easily digested because they are intimately mixed with and are embedded in cellulose cells. Cellulose as is well known, is not easily decomposed by the digestive ferments present in the human body. Cows and buffaloes largely eat and live on cellulose and the time they take to masticate and digest the cellulose is well known. Great care and time are required in cooking vegetable foods because the cellulose walls must be broken up and the protein and other contents of the food must be liberated and made soluble and assimilable. In the case of animal foods, the digestibility and assimilability are greater only if the great danger of fermentation and decomposition that easily set in in the animal food is not taken into account. However, it is to be remembered that although the cellulose is indigestible, its presence under normal conditions in suitable amounts is advantageous. Although it has no nutritive value (as human food), it is supposed to be necessary as a stimulant to intestinal peristalsis by acting as a gentle mechanical irritant, by helping to retain moisture and keep the fæces soft and by giving such bulk that the intestinal muscles can act to good advantage. Indeed according to one school of medical thought, a certain amount of cellulose is absolutely necessary in the human diet. In the case of a non-vegetarian diet where there are no vegetables and where no vegetarian food or grains are included, the food is so fully absorbed that the large intestine has no stimulus to action and no material for its muscles to act upon, and as a consequence, constipation sets in with serious results. It is for this reason that in constipation, agar agar—a non-irritating form of cellulose—added to food has proved medically a successful therapeutic measure.

XVIII

TEA, COFFEE, COCOA AND TOBACCO.

Of the three drinks tea, coffee and cocoa, there is not much misunderstanding regarding the use of cocoa as a valuable drink. This drink is a nutritious one and is valued as such by all countries alike. Regarding the use of tea and coffee, however, there are considerable differences of opinion. In India, particularly, even the elementary preparation and use of tea are very much misunderstood. Whereas there is one class of people that bans the use of tea as being a dangerous drug, there is the other class that is so much given to tea that it lives on practically a decoction of tea all the day long. England in the West and Japan and China in the Far East are examples of nations that make a liberal use of tea and are at the same time having a very healthy race of people. In Japan, morning, afternoon and evening, a plain, mild, watery extract of tea "O-cha" (Honourable Tea, as it is called) is a common drink, but it should be noted that ordinary water is never used for drinking purposes at any time of the day. The mild decoction of tea without milk and sugar serves the purpose of drinking water and at the same time eliminates altogether all the dangers accompanying the use of ordinary water carrying all sorts of bacteria of dangerous diseases as in the case of India. In Germany, as in Japan, hardly any cold water is ever used for drinking purposes. Beer, aerated waters and coffee are used in the place of ordinary drinking water. In France, coffee and the weak wines are used instead of water. Similarly in other European countries, use of water is avoided entirely for drinking purposes. Those of us in India who criticise the use of tea or coffee should bear this important fact in mind that in all the colder countries, whether it be in the Far East like Japan or Russia in the West, where it is nearly impossible to drink cold water, hot and mild drinks like tea and coffee are freely used *as substitutes* for cold water and hardly any cold water is used for drinking purposes. In India, on the contrary, tea and coffee are most unscientifically prepared as heavy decoctions and are used as substitutes for food; and at the same time ordinary water, unfiltered and insanitary, is used both for drinking and cooking purposes with all the dangers accompanying it. If tea and coffee, rightly and correctly prepared, are taken (along with some food) as a substitute for water, they are not likely to do much harm. Indeed, when one knows the extreme difficulty of getting pure milk in the big cities of India, it is perhaps safer to take a cup of hot tea with a little milk in it rather than take a cup of doubtful milk with all the dangers accompanying it. If taken in *moderate* quantities and in the *proper* manner as a substitue for water, no harm should be expected from tea or coffee. As explained above, whole nations are none the worse for these drinks and the lesson to be learnt from them is a simple one.

Tea.

Tea has been used by the Mongolians from very remote ages. According to a Chinese traditional belief, tea is said to originate from the superfluous growth of the eye-lashes of a pious Chinese ascetic (who lived about the year 500 A. D.) by name Darna, who had cut his eye-lashes to remain awake in his penance.

From China, its use spread on to East India and it was introduced in Europe in the seventeenth century. The tea leaves are treated in two different ways and are sold as green tea and black tea. The green tea is lightly roasted and coloured; the black tea is first subjected to a process of fermentation (cured) and then slightly roasted. Some of the blends of black tea leaves sold in India are supposed to be artificially perfumed with small quantities of aromatic oils to give them a special flavour. Teas grown at high altitudes, like the famous Darjeeling tea, have a peculiar flavour of their own due to the presence of certain essential oils.

Dried tea leaves are made up of 3·7% ash, 15% Tannic acid, 1·5 to 2·4% Theine (identical with Caffeine) and 0·6 to 1·0% essential oil. Boiling water should be poured over the tea leaves and allowed to act for a few minutes (not exceeding five) in a closed porcelain vessel. If the tea is allowed to brew for a longer time, the tannic acid gets extracted. This is harmful to the human system.

Caffeine also is extracted along with the tannic acid, and this, in large doses, is certainly harmful ; for only a small quantity of the alkaloid has a stimulating as well as beneficial action on the nervous system.

There are different ways in which tea is taken : in Japan a watery extract is taken without sugar and milk ; in Europe, many take it with a little lemon juice ; in Kashmir, Skardu, Leh and Ladak, it is made into an emulsion with salt and a little milk and sugar.

Coffee.

Coffee is a national drink, in the central states of Europe, particularly Germany, France and Switzerland where its use is encouraged in preference to tea by levying a heavy import duty on the import of tea. England encourages the use of tea in preference to coffee. In India, coffee is a popular drink in the South of India and on the West Coast. Even in the Nilgiris, where high class tea is grown, coffeei s a popular drink. The green coffee seeds are to be roasted at a particular temperature at which they get a nice aroma The average composition of the green coffee seeds is the following :—

10 % Water.

12 % Nitrogenous matter.

1 % Caffeine.

12 % Fat.

12 % Sugar.

9 % Tannins.

38 % Cellulose substance.

5·5 % Ash.

On roasting, the seeds undergo a chemical change. The sugar is naturally caramelised; a sort of dry distillation of the seeds takes place giving rise to a peculiar coffee aroma and a characteristic taste. The fat of the seeds is transformed to a kind of coffee oil and the seeds become brittle and easily grindable. The coffee when taken (like tea) works as a nerve tonic, removes the feeling of tiredness, gives fresh energy for work and reduces sleepiness. Of the two drinks, tea and coffee, tea is said to act on most constitutions as a laxative, whereas it is supposed that the reverse is the effect in the case of coffee.

Strong extracts of both tea and coffee are supposed to make the kidneys move actively. As the food value of coffee is small, in Germany it is taken along with sugar and cream. Coffee is often sold in the market mixed with chicory.

Cocoa.

Cocoa as a drink was known first in Mexico, from where Cortez brought the seeds to Spain in 1520, and then its use spread to the rest of Europe. Cocoa seeds are known to have been used in Ceylon for centuries. The cocoa seeds are first separated from cocoa-fruit by fermenting and are then dried. The dried seeds are then roasted at a temperature below 100° C and are finely powdered. The dried powder contains about 50% fat, known as cocoa-butter which is separated by expression under a hydraulic press. With varying percentages of fat, the cocoa powder as such is sold in the market. In Holland, the cocoa powder is treated with a mild alkali like ammonia, potash or soda carbonate and is then put on the market, which, however, affects the aroma of the cocoa. The cocoa powders available in the market contain from 10 % to 30 % of fat (cocoa-butter). This fat is a very easily assimilable and highly nourishing constituent of cocoa and should not, therefore, be removed from the cocoa-powder. Besides fat, cocoa-powder contains 13-16 % carbohydrates, about 18 % nitrogenous matter and 2 % theobromin. When compared with tea and coffee, cocoa is a nourishing drink and as such produces a feeling of satisfaction. The theobromin produces, like caffeine (in tea and coffee), a refreshing feeling.

Chocolates are prepared by mixing the cocoa-powder (natural, *i. e.*, with all the cocoa-butter in it) with sugar and some spices. In Central Europe, like tea and coffee, "Kakao (cocoa) and Schokolade (chocolate)" are two favourite and nourishing drinks. In the cheap brands of chocolates, starch and sugar are used to lower the cost.

Tobacco.

In the West-Indies and Mexico, tobacco was already in use for eating, inhaling and smoking purposes even before Columbus discovered America. Tobacco seeds were imported into Europe about the middle of the sixteenth century. Up to the end of the sixteenth century, it was used for its medical properties. In the seventeenth century, it began to be used for eating and smoking purposes. Pope Urban banned everyone who inhaled tobacco from entering the Church; Sultan Murad IV passed a law in 1633 punishing every smoker with penalty of death ; and in Russia the punishment for smoking was the cutting off of the nose ! In spite of these punishments, tobacco is getting to be increasingly popular. Indeed in some cases, the sale of tobacco is taken to be an index of civilisation ! Tobacco is one of those substances the use of which can be easily given up. Due to the presence of the alkaloid, nicotin, it stimulates the nervous system, the after-effects of which are undoutedly ruinous to the human system. To many, tobacco smoke is very repulsive and creates a nausea which is intolerable. Even in Europe, where smoking is not considered bad manners, it is interesting to note that in travelling, whole compartments are reserved separately for smokers and non-smokers. In India, no such facilities are afforded to the non-smoking public to protect them from the nuisance of the smoking public. Tobacco has very little to recommend itself and in India, at any rate, the use of this irritating leaf should be controlled by severe and strict legislation, as is done in the case of alcohol.

XIX

ALCOHOLIC DRINKS, THEIR NATURE AND FOOD VALUES.

Alcoholic drinks.

Alcoholic drinks are very much in demand and are very much used, in spite of religious prohibitions, in the case of certain communities. In a certain sense, alcoholic drinks are used as appetisers. In restaurants in Europe, it is a common practice to begin the food with a drink as an appetiser. More than as appetisers, drinks are also in-demand as intoxicants,because they enable a person to drown his misery, tiredness and the day's anxieties in forgetfulness. Man is a rational being, but all the same he is more surrounded with grief than with happiness. He is, therefore, on the look-out for some means by which he will forget the day's grief and misery although the means employed be harmful. It is interesting to note in this connection that the Muhammadan mountain tribes, living in the Caucasus mountains, use a fermented (alcoholic) milk preparation equivalent to a light wine, known as Kefir, although their Prophet has prohibited the use of intoxicants like wine and beer. Certain other races use other intoxicants like opium, charas etc. for similar reasons.

There is not the least doubt that alcoholic drinks are harmful to the human system, in the long run, and more particularly so in tropical countries where, even in small quantities, their use is positively detrimental to the human system. In extremely cold climates, the use of alcoholic drinks in certain limited quantities is justifiable for the quick warmth they impart to the human body but even then, the slow and gradual harm they do to the protoplasm and other organs of the body, more than outweigh the advantages, if they are not judiciously taken. The great danger that accompanies the use of these alcoholic drinks is that they create a force of habit which outbalances the sense of judgment, with disastrous results to the human system that cannot be controlled even by the rational element present in man.

The psychological action of alcoholic drinks on the human body is that they produce heat and spare the use and consumption of fat present in the human system. In this limited sense, alcohol is a nourishing material. Along with this reaction, however, there is the dangerous working of the alcohol on the cell protoplasm, which gradually gets weakened. The wines obtained from potatoes are found to be particularly destructive in their action on cell protoplasm because of the fusel-oil they contain.

It is also an established fact that alcohol produces a slow destructive action on the liver, heart, kidneys and the nervous system. It is difficult to lay down any quantitative relation between the amount of alcohol taken, and the harm it does to the system, because it is dependent on so many factors, both internal and external, such as the constitution of the individual, the surrounding temperature etc.

The alcoholic drinks may be roughly divided into three classes, light wines, beer and spirits like whisky and brandy. The light wines are prepared by the fermentation of grapes and are a very common drink in countries like France and Switzerland where grapes are cultivated extensively. The sugars contained in the grapes are fermented into alcohol and carbon-di-oxide by the ferments present in them. In addition to containing alcohol, these wines contain also acids like acetic acid, citric acid, tartaric acid etc., and other appetisers. These light wines are taken with the usual food almost like water in France. As such, they are not to be considered as nourishing materials, but they are to be considered as condiments and appetisers. Such wines contain as a rule, 9-12% alcohol, 2·0% extract and 0·1-0·8% sugar. A preparation similar to the above is made in India from dried grapes after subjecting it to a process of fermentation. Qualitative and quantitative data regarding the composition of this preparation, known as *Drakshasava* (द्राक्षासव), a valuable tonic according to the Ayurvedic system of medicine, are not available and it is, therefore, difficult to say what percentage of alcohol is present in this preparation.

Another class of wines is the so-called fruit wines. These are prepared from fruit-juices such as apples, pears, berries of all kinds etc. The taste of these wines is different from that of the grape-wines which are supposed to taste much better and are considered more delicious than the former. In these fruit-wines, the percentage of alcohol is less than in grape-wines.

Beer is a drink obtained by fermenting barley and is subjected to a scientific process of fermentation. The German beer is reputed to be one of the best beers in the world and even there, the beer from Munich (Munschener Bier) is reputed to be the most famous. In Germany beer is a national drink and is sold even cheaper than aerated waters. Just as in India, one thinks of cold water for quenching one's thirst, so in Germany the substitute for water is beer. Japan has also begun to manufacture beer as in Germany. The average beer in Germany contains between 2 and 5 % of alcohol and about 5-6 % of extract and is, in a sense, a nourishing food. It should be noted that beer meant for export contains a large percentage of alcohol, as in tropical climates the fermentation goes much higher than in cold climates. Beer is very much less injurious than wines and is certainly better as a nourishing material.

The whiskies and brandies are the most detrimental of all the wines to human health as they contain a very large percentage of alcohol. These are made from potatoes, maize, rice etc., in different countries. The starch present in these corns is first converted into sugars which are then fermented into alcohol. In India, flowers of *Mahuwa* (rich in sugars) and molasses are used for a similar purpose. Rum, arrak, kongnak etc., are different modifications of these alcohols mixed with different flavouring materials and spices. The number and nature of these drinks in the market is almost un-ending as new mixtures are every year being introduced for sale. These wines contain alcohol from 20 % onwards to 55 % in the case of kongnak and 77 % in the case of rum. For the vegetarian in Europe, attempts are being made to replace these alcoholic drinks by non-alcoholic syrups. Although there are some good drinks in the market which are non-alcoholic, yet strangely enough, they are not being manufactured so cheap as the alcoholic drinks.

It should be borne in mind that the alcoholic drinks are unsuited for the hot climate of India. The alcoholic drinks in Europe are safe in the sense that once the fermentation is stopped in the manufacturing process, there is no chance of any further fermentation because they can be stored in cool cellars. In India, the fermentation increases with the temperature with the result that it is not possible to take any of these drinks without positively doing harm to the nervous system. To use alcohol in India as in Europe is nothing short of courting a disaster.

XX.

Milk and Its Medicinal Uses.

The Ayurvedic system of medicine has highly recommended the use of milk both for purposes of curing disease as also for improvement of health. As has been pointed out in a previous chapter, the different types of milk have been examined and their use advocated for different diseases. In one special treatment, however, the exclusive use of milk for curing intestinal diseases forms a unique feature of this system. The treatment is known by the name "*Parpati*" (पर्पटी) which is descriptive of the scaly and veneer-like nature of the special medicinal drug prepared. In difficult diseases like Sprue, chronic Diarrhoea, Dysentery, mucous Diarrhoea etc., the patient is kept entirely on milk diet, so much so that some Ayurvedic physicians go to the extent of totally prohibiting water. The *Parpati* is a compound obtained by the interaction of mercury and sulphur the quantities of which are exactly prescribed for the different varieties of *Parpatis*. In addition to mercury and sulphur, metals like gold, or copper, or iron etc. are added and subjected to a prolonged process of rubbing and mechanical mixing in a stone mortar, thus incorporating these metals in a finely divided form. Each variety of *Parpati* is meant for a particular disease. A certain amount of free sulphur in the *Parpati*, in a finely divided form is particularly recommended for preventing gastric action and for making it effective on the intestines.

Milk combines in itself the properties of a complete food as well as an intestinal antiseptic. From the day the treatment begins, every other kind of food or drink is prohibited, and the patient starts with pure milk and doses of *Parpati* at regular prescribed intervals. Hunger and thirst are both to be satiated by milk only. If the milk is taken in excess, it is supposed to do no harm whatsoever; on the contrary, it is supposed to clean and disinfect the intestines. The *Parpati* is the tonic, the constituent metals of which in their finely divided state accelerate the internal bio-chemical reactions. The digestive organs increase their activity and the intestines are in a sense re-built, besides being disinfected. As a result, the patient gradually recovers and begins to assimilate more and more of the milk. In a sense, the adult is once more treated exactly like a baby living on nothing but milk. It is incumbent on the patient to take absolute rest, mental and physical, also like a baby.

The cures effected by this system, in several cases, have been nothing short of marvellous. Serious cases given up as hopeless, by the best of doctors trained in the allopathic system, have been successfully treated by this method and cases are on record where patients who could not digest even half a pound of milk a day have, after a *Parpati* treatment of about two months, grown so healthy as to assimilate as much as *fifty six pounds* of milk per day. Even with such large quantities of milk, the bowels are known to move absolutely normally. A famous Ayurvedic physician of Benares, of an All-India reputation, Trimbak Shastri, is a specialist in the *Parpati* treatment and has saved many valuable lives by this treatment. The example given above is from his record.

SOME EUROPEAN PROVERBS REGARDING MILK

Proverbs.

From the cradle to the grave, milk is a valuable food.

If you are thirsty, drink milk.

The white milk gives to the children red blood, ruddy cheeks and bright eyes.

Plenty of milk makes house-keeping cheap.

He, who wishes to economise in the household, will do it best with good milk.

Milk is liquid flesh.

To him who works hard, milk gives strength.

Milk is healthy for the young and old, it does not matter whether it is hot or cold.

The beer makes you dull, the wine gives you gout, the brandy gives you copper looks, the port wine thickens the blood, the champagne bends and cracks the bones. What the young and old can ever drink—

that is milk ; it nourishes the man ; it not only freshens the heart, but it keeps the head clear and bones erect.

Milk stands for beauty, good teeth, health, increased efficiency, happiness and long life.

Milk is the cheapest and most nourishing food-stuff available to mankind. Milk builds your muscle and your bones and makes you healthy. A healthy man is made immune against disease and, therefore, milk for a healthy man is protection against disease.

Relatively speaking, milk is also the cheapest of food-stuffs. Therefore, drink milk and use other milk products freely.

Milk contains the energy-giving nutrients, protein, fat and carbohydrates ; all the known essential vitamins ; calcium, phosphorus, iron, sulphur, iodine, magnesium, potassium, sodium, chlorine and copper, some of the physiological roles of which are known ; and a number of other elements, present only in minute amounts, such as manganese, zinc and fluorine, the exact functions of which are not fully understood, but which would seem to be as necessary for normal nutrition as any of the other constituents.

Mussolini and milk.

Mussolini has written as follows in one of his articles entitled "My Twenty-four Hours". He says, "My breakfast consists of a plain glass of milk and I hardly require a minute for taking it. Milk is a wonderful nourishing material, perhaps the very best that is known to man. This one glass of milk contains sufficient nourishment to keep me quite fit for steady work up to my mid-day meals—six hours later."

XXI

BREEDING OR FEEDING?

वत्सो वृषो धेनुश्चैषामवध्याः ।
घातः पंचाशत्को दण्डः ।

"Cattle such as a calf, a bull or a milch cow *shall not be slaughtered*. He who slaughters or tortures them to death *shall be fined fifty Panas*."

(Kautilya's Arthashastra.) 2-26-43.

Although scientific trials on cattle-breeding are being carried on in India, both by Government and private agencies, for the last several years, it must be confessed that from the time Lord Linlithgow has come over to India, as the Viceroy, this question has received a new momentum. As a result of the personal interest he has taken in the matter, the provincial authorities have been bestowing special attention upon this question. There is a new consciousness, as it were, all over the country, for the improvement in the breed of cattle. There is not the least doubt that this move is in the right direction, but it is a question how far the improvement of the *breed alone* will solve the problem of increasing the total supply of milk. It is true, as the saying goes, good breed is half the herd but the real difficulty is that without an adequate provision for the feeding of the improved animals, the main problem will remain unsolved. At the very best, it is like sowing very healthy seed on a soil which is lacking in proper irrigation and suitable fertilisers. The new campaign of presenting or distributing stud bulls is, therefore, like scattering fine seed on a poor soil. In India, today, agricultural experts are agreed that, in most provinces, the breeds of the cows or buffaloes are not so poor as they are made out to be. According to recent researches carried out by Government experts, it has been shown that cattle fed on proper nourishing food have

IF

YOU

WANT

MORE

MILK

GIVE ME

SUFFICIENT NOURISHING FOOD.

Milk is Seven Times as Nourishing as Beer!

yielded easily fifty per cent. more milk than otherwise. Breeding apart, if the millions of cows and buffaloes are properly fed, the milk supply of India will go up by millions of gallons annually. As important as the question of breeding, therefore, is the equally important question of adequate feeding.

The serious question before the country, therefore, is this. Is the Indian farmer economically in a sound position to be able to purchase the necessary food-stuffs to feed his cattle ? Is it the want of scientific knowledge and his ignorance that stand in the way of his paying the necessary attention to his cattle ? Is it his economic helplessness that is really responsible for this deplorable state of affairs ? The agriculturist in India is uncharitably criticised as being ignorant and conservative. He is being looked upon as callous and unkind to his cattle. He is being told, he does not realise the importance of good fodder and good manure. It is worth the while examining some of these charges.

As to the charge of illiteracy, the total percentage of literacy in the country easily proves this charge. But who is responsible for this appalling illiteracy in the country? Certainly, not the farmer, who is paying taxes to the state more than his income can justify his doing so. The charge of his being conservative also does not stand, because those who have studied the question know that he is progressive so far as his economic condition permits him to be. The price of the land on which he is toiling from morning till evening and also the cost of the produce he is raising from a soil whose fertility is being impoverished year after year, have fallen in value by fifty per cent. The burden of taxation on the land has not been correspondingly lowered. The little gold he had with him as his saving is deserting him week after week as " distres gold". He is selling all he has and all he can only to exist on his land. The economic surveys made in different parts of India have shown that the farmer is everywhere submeregd in debt. He has stuck to his land because he does not know to which other planet he could migrate ! He loves his cattle and knows the importance of feeding them. He also understands the meaning of cattle food and oil cakes but he has nothing in his pocket to pay for these in exchange. He realizes the meaning of cattle dung as manure which he has used for generations as a fertiliser. But, he has no money to buy the necessary fuel to warm up his half-clad body on wintry days and he has to burn the nitrogenous fuel to cook the food, only to keep his body and soul together. The pastures that were once open to his cattle for grazing are now closed. No wonder then that the farmer should feel that those that try to preach to him on these topics are offering him gratuitous advice.

In the new campaign, the villages are being presented with stud bulls and it is suggested, "something" has got to be arranged for their up-keep and feeding. Who is going to be responsible to see that this "something" is arranged for ? Surely, the farmer who is not able to look after himself or his own cattle is not to be expected to shoulder this burden however much it be for his own good ! The real remedy is to provide new pastures round the villages and reserve areas for the feeding of the cattle. A year of famine means to the farmer to-day absolute starvation for himself and for his dear cattle which (latter) he has to sell away to the slaughter houses. Indeed, as a great authority on the subject has put it, the introduction of the institution of slaughter houses in India is synonymous with the beginning of famines ! In ancient India, in the days of Kautilya, the slaughter of cattle (as given at the beginning of this chapter) was prohibited by law under penalty of a heavy fine.

The number of milch cattle in India year after year is either remaining steady or is going down. The yield of milk per (milch) animal is also decreasing. The exports of hides and skins have their own story to tell. The slaughter houses are not having any slack business : every drought and famine of fodder adds to their prosperity. The figures for the export of oil cakes which are used by the importing countries as cattle food and manures also point to the poverty of the Indian farmer who cannot afford to retain them for the use of his own land or for his own cattle. The following are the figures for the last three years showing the exports of oil cakes and hides and skins raw and tanned.

	Total oil cakes. (In Tons.)	Value in Rs.
1932-33	2,89,701	1,78,49,859
1933-34	3,50,573	1,92,68,074
1934-35	2,89,604	1,75,97,270

	Total Hides and Skins Raw (in Tons.)	Value in Rs.
1932-33	38,129	4,06,30,462
1933-34	39,811	3,26,47,824
1934-35	47,230	3,95,93,122

	Total Leather (in Tons.)	Value in Rs.
1932-33	17,555	5,06,76,099
1933-34	18,476	5,36,88,634
1934-35	19,654	5,17,98,795

From the above, it will be clear that mere improvement in breeding alone will not be sufficient to solve the problem of milk supply in India. The population of India is increasing by 3 millions every five years thus contributing to an increased scarcity year after year. Breedng and feeding must go hand in hand if the healthy cattle produced have to be of any use either as milch cattle or as beasts of burden. Indeed, without an adequate provision for feeding, more harm is likely to ensue than good. One cow well-fed is more fruitful than five cows ill-fed. Without a radical change in the economic status of the agriculturist, breeding alone will be like giving systematic exercise to the body without making any provision for its nourishment.

The following extract explaining the typical prosperity of a small country like Denmark should serve as a model to India.

The transformation which occurred in Denmark, (a country of half the size of Mysore state in India !) from a country still exporting cereals in 1880 into a country with a large export trade in milk products and meat, is an example which merits careful examination. This transformation has made necessary a great increase in the supply of feeding stuffs for cattle, a demand which has been met *by* an *ever increasing importation* of *cereals* and *oil-cake* accompanied by a marked intensification of domestic cultivation.

The area devoted to cereals remains to-day, as in 1880, 41% to 42% of the agricultural area ; but the area under root crops has risen from less than ½% to 14% of the agricultural area, while the area under fallow has almost disappeared.

The exportation of agricultural products has developed as follows ;

....	Butter (Millions of Kilos).	Condensed Milk and Cream (Millions of Kilos.)	Bacon (Millions of Kilos.)	Eggs (Millions of Kilos.)
1881-1885 ..	14	..	8	60
1931 ..	172	19	376	980

XXII

Cereals, Pulses, Fruits and Vegetables.

The vegetarian food stuffs available in India can be broadly classified under three headings, (*a*) cereal grains and pulses, (*b*) vegetables and (*c*) fruits. The cereal grains and pulses vary from province to province. Whereas, rice, for instance, forms the staple food of the Bengali and the Madrasi, it is looked upon as a food fit for patients in the Punjab. Wheat and pulses, rich in nitrogenous matter, are so commonly used in the Punjab by the rich and the poor alike, but in South India wheat preparations are reserved for holy holidays only. Bajra and Jowar are very common substitutes (for wheat) in the Deccan plateau and Ragi takes its place in the Mysore State.

In the case of vegetables, also, an equally great diversity is to be found throughout the length and breadth of India. In the Gangetic valley and in the Punjab, some of the best and cheapest vegetables unknown to Europe are grown in plenty. Some of these cannot be had in the rest of India, particularly in South India, either for love or for money. In the Himalayan valleys and the important hill stations, vegetables are a luxury being all transported from the plains and sold at almost prohibitive rates. In the Deccan plateau and the Madras Presidency, the vegetables grown are comparatively poor in quality when compared with those grown in North India and are not cheap either. In Europe, potato is a cheap and common vegetable of almost every day necessity whereas in India, people go without it for months together. The tomato is being grown practically all over India.

Regardnig fruits, the distribution of these all over India is an interesting study. The Himalayan valleys and Kashmir produce a large quantity of apples, sweet and sour, and yet the communications in the hills and the railway facilities are so inefficient that, like foreign coal which is sold in India in competition with the Indian, foreign apples are being imported from Japan and America and sold in the Indian market ! The banana is largely grown in the Bombay and Bengal Presidencies and cannot be had cheap in the Punjab or the United Provinces. The oranges from the Central Provinces and Assam are famous all over India. The figs of Poona are a speciality and are equally famous. Very few of the Indian Provinces excepting the Bombay Presidency can enjoy the fresh fig. Baluchistan, Kashmir and Kullu are sending out to the rest of India their grapes, peaches, almonds and apricots. The mango is a valuable and delicious fruit and is grown in the different provinces of India with its characteristic flavours peculiar to each province. The railways, in India, have scarcely paid any serious attention to the question of transport and distribution of fruits by devising anything like special cold storage vans.*

It should be remembered that fruits and vegetables play a very impotant role in the adjustment of a well-balanced diet in India and deserve more careful attention than has been bestowed on the subject so far. The following tables are, therefore, of special value not only because of the very wide field they cover but also because the analytical data refer mostly to Indian samples.

---

*Only recently, some railways in India have begun to pay some attention to this question.

CEREALS.—Jowar (Bot. Andropogon Sorghum.)

Name.	Moisture.	Oils & fats.	(N×6·25) Proteins.	Carbo-hydrates.	Ash.	Cellulose. Wood-fibre.
Nagpur	10·0	7·7	7·6	70·1	2·3	2·2
Punjab Br. ..	11·3	3·2	8·9	71·9	2·6	2·1
Punjab-White ...	12·0	3·1	8·2	73·5	1·8	1·4
Cawnpore ,, ...	11·4	2·6	8·2	73·5	1·7	1·6
Shaler	6·2	0·6	11·8	77·6	1·6	2·2
Utavli	4·9	2·8	12·7	76·3	2·0	1·3
Lalbondi	6·2	3·3	10·0	77·4	1·7	1·4
American-Sorghum ..	5·8	3·4	9·5	73·8	3·1	4·4
Maize (Bot. Zea Mays).						
Maize Poona ...	5·6	3·7	8·9	75·6	2·5	3·7
,, Punjab-White ..	11·9	4·5	8·3	72·0	2·1	1·2
Maize, King Philip Cawnp.	10·7	4·2	11·6	70·3	1·7	1·5

CEREALS (*Contd.*)—**Average and General Composition of Wheat** (Bot. Triticum Sativum).

Name.	Moisture.	Oils & fats.	(N×6·25) Proteins.	Carbo-hydrates.	Ash.	Cellulose. Wood-fibre.	
W. Khandesh ..	11·7	2·0	16·7	66·5	1·7	1·4	
S. P. Bukshi ...	12·5	1·8	17·6	65·1	1·7	1·3	
Banshi	11·1	2·3	16·5	66·4	1·7	2·0	
Austr. Hybrid ..	12·1	1·8	15·2	67·4	1·7	1·8	
Bombay Pre. ...	8·8 to 12·5	1·4-2·7	14·0-18·8	63·0-72·1	1·3-2·1	1·1-2·4	Average.
Rust. Pre. Cawnpore ..	9·9	1·5	9·3	75·9	1·9	1·5	
Cawnp. Beardless ..	13·2	1·6	9·8	72·0	1·5	1·9	
„ Bearded ...	13·4	1·7	8·5	73·0	1·8	1·6	
England	..	1·86	10·99	69·21	1·67	2·9	Average.
India	..	2·08	10·99	70·90	1·45	1·92	Average.
Europe (Russia excluded)	13·7	1·8	12·3	67·9	1·80	2·5	Average of 208 Samples.
American	10·2	2·2	12·2	71·7	1·9	1·8	Average of 407 Samples.
All Countries ..	13·3	1·85	12·03	68·67	1·77	2·31	

Cereals (*continued*).

Composition of some varieties of Indian Rice. (Oryza Sativa.).

Name.	Moisture.	Oils & fats.	(N×6·25) Proteins.	Carbo-hydrates.	Ash.	Cellulose. Wood-fibre.
Ambemohor (Poona) ..	11·5	0·8	6·5	80·6	0·6	Nil
Kamod (Nasik) ..	7·5	1·2	6·5	84·0	0·7	0·10
Dilmohor (Konkan) ...	11·5	0·7	6·1	81·3	0·4	Nil
Ambemohor („) ...	11·5	0·7	6·2	81·1	0·5	Nil
Kolumba (Karjat) ..	8·3	1·0	6·2	83·8	0·5	0·2
Ambemohor (Poona) ...	7·7	1·05	6·8	83·7	0·7	0·05
Bezwada ...	10·4	0·8	6·7	81·5	0·6	Nil
Rangoon M. ..	8·2	1·3	7·2	82·5	0·8	Nil
Rangoon Nani ...	6·0	1·0	7·1	85·0	0·8	0·1
Bengal Winter ..	12·5	0·9	6·4	79·2	0·8	0·2
Assam ...	12·6	1·7	6·4	78·6	0·8	0·25
Average of a number of samples.	5·98—11·5	0·65—1·77	5·18—17·5	86·65—85·06	0·35—0·8	0—0·2

Cereals (*continued*).

Bajri or Bajra (Eng. Bulrush millet). (Bot. Pennisetum typhoidenon).

Name.	Moisture.	Fats & Oils.	(N×6·25) Proteins.	Carbo-hydrates.	Ash.	Cellulose Wood-fibre.	
Ahmadnagar ..	6·7	5·6	12·9	73·5	1·8	0·5	
Poona ..	7·5	5·2	9·8	74·6	2·3	0·6	
Gujrat Pranty ..	6·6	5·0	10·9	75·1	2·4	..	
Gujrat Daskroi ..	6·21	5·5	11·7	74·1	1·9	0·6	
Punjab ..	10·1	4·9	11·1	71·1	2·0	0·8	
Nagli, Ragi, or Bavto (Eleusine Corcana) : Inferior Millets.							
Ragi, (Poona) ..	14·4	1·3	6·5	73·3	2·7	1·8	
Ragi, Mysore ..	13·2	1·2	5·4	75·1	3·0	2·1	
Vari, Poona ..	8·0	4·1	6·8	67·3	6·2	7·6	Panicum miliaceum
Kodra or Harik, Poona	8·0	3·4	5·8	70·1	4·3	8·4	Paspalum
Kodra or Harik, Ratnagiri	9·1	3·4	5·5	70·8	1·4	9·4	Scrobiculatum

Cereals—(*continued*).

Name.	Moisture.	Oils & fats.	N×6·25 Proteins.	Carbo-hyd-rates.	Ash.	Cellulose. Wood-fibre.	
Rajgira (Amarantus paniculatus).							
Rajgira Poona ..	8·9	5·3	15·4	65·8	2·7	1·9	
Pulse Grains. (Leguminous Plants).							
Eng. Pigeon Pea : *Tur*. (Bot.—Cajanus Indicus.)							
Deccan ..	6·0	1·6	21·1	65·8	3·3	1·1	Split
Gujarat ..	2·5	1·4	22·9	69·3	2·9	1·0	Split
Cawnpore ..	6.6	1·6	20·0	67·4	3·0	1·4	Split
Un-Split Grains.							
Deccan ..	6·9	2·5	19·6	60·8	3·5	6·7	
Gujarat ..	6·3	1·5	20·7	60·8	4·3	6·4	
Cawnpore Seed, White	10·8	1·5	14·2	63·7	3·5	6·2	
,, Seed Red ..	10·9	1·0	16·6	62·9	3·7	4·8	

Pulse Grains.—(*Continued.*)

Name.	Moisture.	Oils and fats.	N×6·25. Proteins.	Carbo-hydrates.	Ash.	Cellulose. Wood-fibre	Bot. Name.
			Gram. (Bot. Cicer Arietinum).				
Average of a number of	7·38	4·11	19·44	52·26	2·70	2·25	
Samples ..	to 9·80	to 6·10	to 20·69	to 59·41	to 3·25	to 9·59	
Gram Kabuli, Cawnpore	10·4	4·2	22·8	57·5	3·2	1·9	
			Udid (Black Gram).				
Udid ..	6·05	1·25	19·81	50·05	3·45	4·25	Phaseolus
	to 11·95	to 2·6	to 27·50	to 60·69	to 5·35	to 5·9	Radiatus
Guar ..	7·60	2·65	27·43	42·60	3·50	7·00	Cyamopsis
Cluster bean ..	to 15·25	to 3·60	to 30·37	to 47·67	to 4·65	to 8·20	Psoralioides
Matki ..	4·60	0·65	22·56	58·49	3·70	4·30	Phaseolus
Kidney bean ..	to 8·15	to 1·75	to 25·50	to 63·20	to 6·30	to 5·45	Aconitifolius
Kulthi ..	4·30	0·65	20·75	56·04	4·20	4·85	Dolichos
Horse Gram ..	to 10·25	to 1·84	to 22·25	to 63·20	to 7·45	to 5·50	Biflorus

Pulse Grains.—(*concluded.*)

Name.	Moisture.	Oils and facts	N×6·25 Proteins.	Corbo-hydrates.	Ash.	Cellulose. Wood-fibre	Bot. Name.
Masur (lentil) ..	8·15	1·75	25·50	63·2	6·3	5·5	Ervum Lens
White Val ..	9·5	2·0	23·4	53·3	4·3	7·4	Dolichos Lablab
Val, *kadwa* ..	9·1	1·1	20·75	58·4	3·9	6·8	Dolichos Lablab

Green and leafy vegetables.

"Green leafy vegetables are rich in the B vitamins and in vitamin A; they are also among the richest sources of the anti scorbutic vitamin C. Since this vitamin is sensitive to heat, the dietary value of uncooked salads is obvious.

The abundance of minerals and vitamins in green vegetables and the special character of the protein, although present in small quantities, make them of great value, especially as supplementary foods to a diet containing cereals . Just as among pastoral communities the defects of a cereal diet may be corrected with milk and dairy products, so, under other circumstances, correction can be obtained by green vegetables. In the densely populated regions of southern China, for example, where animal foods cannot be raised, the deficiencies of the rice diet are largely repaired by the abundance of vegetables eaten."

Vegetables.

Name (Marathi).	Botanical name.	Moisture. %	Fats.	Proteins.	Carbo-hydrates.	Cellulose.	Ash.	Name Hindi.
	N. O. Amaranthacæ.							
Red माठ ..	Amarantus blitum ..	84·0	4·1	18·7	50·6	7·3	19·3-0·81 SiO_2	चौराई
Rajgira ..	Amar-paniculatus ..	80·0	2·7	17·8	50·7	10·4	10·4-0·80 SiO_2	रामदाना का साग
	N. O. Cheno-podeacæ.							
Palak ..	Spinicea obracea ..	84·0	6·25	0·75	63·88	9·12	20·0-0·75 SiO_2	पालक
Chandan-Batwa..	Chenopodium Amb-rasioides.	86·6	5·14	18·18	59·23	7·31	10·14-2·61 SiO_2	
	N. O. Compositae.							
Kardi ..	Carthamus tinctorius.	86·0	6·14	28·12	44·46	9·14	12·14-0·43 SiO_2	कुसुम का साग

Vegetables—(*continued.*)

Name (Marathi).	Botanical name.	Moisture %	Fats.	Proteins.	Carbo-hydrates.	Cellulose.	Ash.	Name Hindi
	N. O. Leguminoseæ.							
Methi ..	Trigonella Foenum græcum.	77·0	4·8	16·21	56·11	11·51	11·37 0·93 Sio$_2$	मेथी
Chuka ..	Rumex Vesicarius ..	92·0	4·62	16·27	57·86	10·5	10·75- 0·75 Sio$_2$	खट्टा पालक
	N. O. Umbelliferæ.							
Kothimbir.	Coriandrum Sativum.	84	3·12	24·46	43·30	9·75	19·37- 1·50 Sio$_2$	धनिया
	N. O. Cucurbitacæ.							
Bhopla (red) ..	Cucurbita maxima ...	89·50	1·00	6·12	77·33	8·55	7·0- 0·17 Sio$_2$	कद्दू
Dudhya (Bhopla).	Vangeria vulgaris ...	90·36	1·24	0·87	75·28	18·05	4·56- 0·21 Sio$_2$	लौकी
Padwal ..	Trico-Santhes anguina.	95.0	2·21	13·75	67·85	10·60	5·6- nil	चिचिंढा
Karela ..	Momordica Charantia.	88·75	2·93	1·62	85·41	1·51	8·53- 0·17 Sio$_2$	करेला
Dodka ..	Luffa Acutangula ...	91·97	2·98	0·87	73·47	16·56	6·12- 0·71 Sio$_2$	तरोई

Vegetables—(*continued.*)

Name (Marathi).	Botanical name.	Moisture %	Fats.	Proteins.	Carbo-hydrates.	Cellulose.	Ash.	Name Hindi.
	N. O. Solanaceæ.							
Brinjal वांगी seed-less.	Solanum Melongena ..	90·80	8·10	16·44	58·36	9·6	7·5- 0·9 SiO_2	बैगन या भंटा
Brinjal (Seeded Variety.)	Solanum Melongena ..	88·26	4·2	16·37	55·23	17·0	7·2- 0·70 SiO_2	
	N. O. Leguminoseæ.							
Guar Green ...	Cyamopsis Psoralioides.	86·0	1·28	21·43	52·29	15·0	10·0	ग्वालीन
Peas Seed Green.	Pisum Safivum ..	77·0	1·00	26·31	57·61	9·56	5·52	सेम
	N. O. Malvaceæ.							
French beans Green.	Phaseolus vulgaris ..	95·0	2·0	23·75	40·25	22·0	12·0	
Bhendi ...	Hibiscus esculentus ..	89·8	1·76	20·18	62·77	7·55	7·74 0·19 SiO_2	भिण्डी

Vegetables—(*continued*).

Name (Marathi).	Botanical name.	Moisture. %	Fats	Proteins.	Carbo-hydrates.	Cellulose.	Ash.	Name Hindi.
	N. O. Cruciferæ.							
Cabbage ..	Brassica Oleracea ..	92·0	3·0	19·50	61·38	8·87	7·25 0·12 SiO_2	पातगोभी
Cauliflower ..	Brassica oteracea Var. Botrytis	90·0	3·30	36·40	41·30	10·50	8·50 (0·10) SiO_2	फूलगोभी
Knol Khol ..	Vari. Caulo-carpa ..	92·8	3·19	27·75	47·12	9·30	12·64 (0·14) SiO_2	गांठगोभी
Mula (Indian Radish.)	Raphanus Sativus ..	91·0	4·00	18·00	52·66	9·34	16·00 (0·33) SiO_2	मुरई or मूली
	N. O. Umbelliferæ.							
Carrots Yellow ..	Daucus Corota ...	81·4	1·72	7·63	74·96	6·56	9·08 0·48 SiO_2	गाजर
„ Red ..	Daucus Carota ..	77·86	1·12	7·05	73·6	10·52	7·71 (0.22) SiO_2	
	N.O. Dioscoraceae.							
Yam ..	Dios-corea Alata ...	73·0	0·44	7·85	86·19	1·48	4·04 0·47 SiO_2	

Vegetables—(*continaed.*)

Name (Marathi).	Botanical name.	Moisture %	Fats.	Proteins.	Carbo-hydrates.	Cellulose.	Ash.	Name Hindi.
	N. O. Aroideæ.							
Suran ..	Amorphophailus Companulatus	78·00	0·50	12·18	76·28	4·00	7·04 0·15 SiO_2	सूरन
	N. O. Convolvulaceæ.							
Sweet Potatoes .. (रताळ)	Ipomea vatatas ...	68·00	4·5	21·45	69·18	1·72	3·12 0·12 SiO_2	शकरकंद
	N. O. Solanaceæ.							
Potato (Deccan).. बटाटा	Solanum tuberosum...	80·66	0·77	16·75	73·58	2·93	5·97-nil.	आलू
	N. O. Liliaceæ.							
Onion (Red) .. (कांदा)	Alluim Cepa ..	85·6	2·17	11·62	78·53	4·02	3·66 0·63 SiO_2	प्याज

Vegetables—(*concluded*).

% of Phosphoric Acid, calculated on dry matter in some Indian vegetables—estimated as $P_2 O_5$.

Name (Marathi).	Bot-Name.	Moisture %	P-2 50 (Phosphoric acid) %.
Palak	Spinacea oleracea	84·0	1·50
Chandan-batwa	Chenopodium Ambrosiodes	86·59	1·76
Methi	Trigonella Foenum Graecum	77·50	1·09
Kardi	Carthamus tinctorius	86·00	1·76
Red Math	Amarantus blitum	84·08	2·02
Pokla	Amarantus mungostannus	82·00	1·79
Rajgira	Amarantus Paniculatus	80·00	1·70
Kothimbir (Coriander).	Coriandrum Sativum	84·00	1·45
Chuka	Rumex vesicarius..	92·50	1·29

Vegetables—(*continued.*)

% of Phosphoric Acid, calculated on dry matter in some Indian vegetables—estimated as $P_2 O_5$.—

Name (Marathi).	Bot-Name.	Moisture %	P-2 50 (Phosphoric acid) %.
Dudhya Bhopla....	Lagneria vulgaris..	90·36	0·85
Red Bhopla	Cucurbita maxima	94·82	2·22
Karela	Momordica Charantia	88·77	1·95
Dodka	Luffa Acutangula	91·97	1·63
Carrots	Daucus Carota	81·4	0·87
Radish	Raphanus Satirus..	91·00	1·32
Suran	Amorphophallus Companulatus ..	78·00	0·69
Yam	Dioscorea Alata	73·00	0·53

FRUIT.

" The special nutritive value of fruit depends on high vitamin C (ascorbic acid) content. Different fruits vary greatly in their vitamin C content, the richest being citrus fruits (*e.g.*, oranges and lemons). Tomatoes are also valuable : grapes are relatively poor.

The total absence of vitamin C from the diet results in scurvy. In this serious condition, haemorrhages may occur in all parts of the body, often accompanied by intense pain ; the gums are swollen and the teeth loosened. With infants and children, serious lesions occur in the bony tissues. Florid scurvy is developed under conditions of severe dietectic restriction ; but there is evidence that a less severe deficiency may lead to a departure from good health, the cause of which is not easily diagnosed.

Vitamin C is readily destroyed by heat, so that fruit jams and preserves usually contain none. It is probable that preserved fruits, however, may possess a dietary value due to their mineral content. Oranges, again, are rich in calcium salts, while grapes are rich in iron, and fruits generally may have a special value in supplying mineral constituents to those on otherwise restricted diets. Fruits having a yellow colour—*e.g.*, oranges and tomatoes—are often useful sources of Vitamin A owing to their carotene content."

FRUITS.

Pomegranates (Bot. Punica granatum) on juice%.

Name.	Non-edible.		Seed only.		Skin.	Juice.	Acidity as (H_2 So_4)	Reducing Sugar.	Non-red, sugar	Total.
Ordinary Good Poona ..	49·4	=	16·80	+	32·6	50·60	0·516	14·56	*nil*	14·56.
Muskat	30·26	=	16·71	+	13·55	69·74	0·27	11·32	*nil*	11·32.

General variation—(average of one dozen Samples) on juice%.

	Non-edible (rind+seed.)	Seed only	Juice.	Acidity as (H_2 So_4)	Reducing Sugars.	Non-red.	Total.
Average	28·63 to 49·4	10·10 to 16·80	57·47 to 71·37	0·37 to 0·78	5·11 to 14·56	*nil.*	5·11 to 14·56

Grapes (Bot. Vitis Vinifera.) %

Name (Marathi).	Moisture.	Ash.	Acidity (as H_2 So_4)	Reducing Sugars.	Total.
Common Bhokari (average) ...	72·8 to 77·2	0·36 to 0·64	0·23 to 0·53	15·69 to to 18·60	15·69 to 18·60

Grapes—(*Continued*).

Name (Marathi).	Moisture.	Ash.	Acidity as (H_2SO_4)	Reducing Sugars.	Total.
	Analysed only for (reducing) Sugar contents %.				
Phakadi	..	..	..	16·4	
Pandhari	..	...	...	18·09	
Bhokari	..	...	..	18·60	
Black Prince	..	..	..	17·10	
Kandahari	..	..	..	19·70	
Kali Sahebi	...	..	...	22·00	

Guavas (Bot. Psidium guajava)

Name.	Moisture.	Total Sugar (on dry matter).	Reducing.	Non-Reducing.
With seed removed:				
Guava fruits, general	76·82 to 87·83	18·27 to 29·18	8·85 to 12·64	8·15 to 15·65
Guava ripened on trees	78·5 to 81·4	22·02 to 30·10	5·48 to 13·96	9·54 to 20·10

Guavas (Bot. Psidium guajava).—(*Continued.*)

Name.	Moisture.	Total Sugar (On dry matter).	Reducing.	Non-reducing
		With seed removed.		
Kothrud	75·4	19·17	8·81	10·36
Dharwar	76.76	28·50	10·04	18·46
Miraj	77·46	29·61	12·31	17·30
Dholka White	82·0	19·55	7·61	11·94
Dholka Red	75·77	17·37	7·0	10·37

Figs. (Bot. Ficus Carica).

Analysis of Figs grown in Poona district.

Name.	Moisture.	Total Sugar (on dry matter.)	Non-reducing.	Reducing.
Figs, February Crop	74·5 to 81·8	38·31 to 51·43	2·89 to 5·76	34·64 to 51·42
Figs, March Crop	69·1 to 81·0	*45·21 to 64·33	3·62 to 15·65	33·15 to 58·63

Figs. (Boi. Ficus Carica).—(*Continued.*)

Name.	Moisture.	Total Sugar (on dry matter.)	Non-reducing.	Reducing.
Figs, April Crop	74·63 to 87·18	*31·01 to 60·18	2·66 to 8·2	29·07 to 54·16
Figs, Ripe completely on plant	47·80	*59·45	3·72	55·73
Figs, Unripe..	83·37	8·76	0·85	7·91
Figs, fully developed but not ripe.. ..	87·8	48·2	*nil.*	48·2

Lemons. (Citrus Family).

Name.	Moisture.	Juice.	Total Sugar.	Reducing.	Non-Reducing.	Acidity.
Lemon (Citrus Madica) Variety Acida.				On Juice.		
Kagadi Lemon ..	77·50	57·60	0·49	0·20	0·24	as { 7·50 Citric acid; 5·25 $H_2 So_4$ }
Sour Lime ..	76·00	56·80	0·36	0·18	0·18	{ 7·74 Citric acid; 5·42 $H_2 So_4$ }

Citrons. (Citrus Family).

Name.	Moisture.	Juice.	Total Sugar.	Reducing.	Non-Reducing.	Acidity.
				(on juice.)		
Sind Lime ..	86·05	39·73	1·38	0·86	0·52	10·7 Citric acid; 7·5 $H_2 So_4$
Long Lime ..	84·57	51·82	Nil	Traces	Nil	10·88 Citric acid; 7·62 $H_2 So_4$
Pomelo (Bot. Citrus Decumana).						
					On Juice.	
Sample I ..	82·56	25·86	4·56	2·79	1·77	1·1 Citric acid; 0·81 $H_2 So_4$
Sample II ..	90·00	..	5·44	2·22	3·22	..
Mosambe (Calculated on edible part). Mosambe (Bot. Citrus Aurantium).						
	88·00	..	13·74	4·16	9·58	

Oranges. (Bot. Citrus Aurantium).

Nagpur Oranges.

Name.	Rind and Seed (non-edible).	Juice	Total Sug.	Redu.	Non-reducing	Acidity
		On Juice.				
Nagpur Oranges ..	38·00-44·00	52·0 to 62·0	7·35-7·90	3·42 to 4·30	3·46 to 3·96	0·81 to 1·21 Citric. 0·56 to 0·85 $H_2 So_4$
	Moisture	Pulp				
Poona Oranges .	82·59 to 85·97	62·22 to 73·33	21·77 to 34·25	11·62 to 13·32	10·15 to 20·93	35·25 to 42.69 Citric 24·67 to 29·88 $H_2 So_4$

Jack-Fruit (Bot. Arto-carpus integrifolia).

	Moisture.	Fats.	Proteins.	Digest. Carbohy.	Fibre (Cellulose).	Ash
Unripe Fruit—Material used as Vegetable in Indian dishes.	64·26	0·86	3·75	26·21	3·30	1·60
Yellow pulp edible	69·20	0·28	2·25	26·08	0·58	0·11
Ripe-fruit—Inner Seed	42·32	0·44	7·19	41·95	1·50	1·60

Mangoes. (Bot. Mangifera India).

Name (On Pulp.)	Moisture.	Total Sug.	Reducing.	Non-Reducing.	Acidity as $H_2 So_4$
Ordinary mango Fruit (Ripe) ..	81·3	58·22	14·46	33·78	0·20
Pairi Mango-Fruit (Ripe) ..	79·3	61·67	9·00	52·69	0·34
Alphonso Mango Fruit (Ripe) ..	80·5	50·67	3·66	47·07	..

(On Pulp) Pairi and Alphonso. (Complete Analysis). %

Name.	Moisture.	Fats.	Woody fibre.	Carbo-hydrates.	Proteins.	Ash.	Sugars.
Pairi ..	84	0.27	0·34	14·04	0·87	0·48	10·08
Alphonso ..	79·0	0·32	0·46	20·77	0·94	0·57	14·36

On Pulp. One Sample.

Moisture.	Potash.	Phosphoric acid.	CaO.	MgO.
83·00	0·15	0·05	0·12	0·10

Custard Apple. (Anona Sp.)

Name.		Moisture.	Reducing Sugar.	Non-reducing Sugar.	Total.
Sitafal (Anona Squamosa)		64·62	5·68	0·87	6·55
Ramfal (Bullock's heart)	Big	61·67	31·47	Nil	31·47
Anona reticulata	Small	64·33	29·30	Nil	29·30

Jambul.—Bot. Eugenia Jambolana. जामून

Jambul Edible matter 68·00 % (On edible matter)	..	8·09	9·26	17·35	Acidity as $H_2 So_4$. 1.21

Figs Dry. (Dry Fruit).

	Persian.	Persian.	Afghanistan.	Grecian.	Smyrna.	California.	Poona.
Moisture % ..	19·45	19·9	19·04	19·14	18·25	19·92	19·95
Sugars % ..	46·30	45·7	46·64	46·50	57·34	57·34	45·95

Dry Fruit. (Dates).

Name. (Dates.)	Seed.	Edible matter.	Moisture.	Woody fibre.	Reducing Sugar.	Non-reducing Sugar.	Total Sug.
Black ..	7·70	82·30	25·00	2·34	59·16	4·20	63·36
Red	13·33	86·67	60·00	1·90	19·38	Nil	19·38
Black tops ..	22·22	77·78	35·00	2·22	46·27	Nil	46·27
Yellow dates ...	20·45	79·55	55·00	2·50	16·23	Nil	16·23
Long tapering ..	11·36	88·64	40·00	2·14	44·07	1·04	45·11
Muskat ...	7·5	92·5	22·10	1·82	76·00	Nil	76·00
Crown ...	7·85	92·15	23·14	1·88	69·8	Nil	69·8

Banana. Bot. Musa Paradisiaca.

Name. मराठी	Skin.	Edible matter.	On Edible matter %	Moisture.	Reducing	Non.reducing.	Total sugars.
Partoli	11·30	..	..	64	9·74	12·9	22·64
Bichirbali	18·10	81·90	..	68·00	3·86	17·81	21·67
Dwarf	17·16	82·82	..	64	7·08	18·30	26·18
Sahasrafani ...	10·10	89·9	..	66·66	9·26	12·74	22·00

Bananas with low percentage of Sugars.

Name. मराठी	Skin.	Edible matter.	On Edible matter %	Moisture.	Reducing.	Non-reducing.	Total Sugars.
Patta modarungbali ..	27·4	72.6	..	72·00	4·31	6·99	11·30
Laturdanbali ..	25·7	74·3	..	70·00	4·32	7·47	11·69
Annebali	25·30	74·70	..	80·00	4·08	6·74	10·82
Mysorebali	13·60	86·34	..	66·00	16·00	3·59	19·59
Yellahibali	12·43	87·57	..	68·00	10·00	9·20	19·20
Raskel	13·10	86·90	..	68·00	11·50	8·00	19·50
Poona	20·30	79·70	..	72·00	4·58	10·49	15·07
Basarai	32·55	67·45	..	74·00	5·32	10·71	16·03
Junnar	27·12	72·88	..	82·00	5·53	7·67	13·20
Bangali	17·05	82·95	..	77·08	8·06	4·04	12·50
Rakhadi	30·8	69·2	..	65·00	2·38	10·24	12·62
Sonkeli	14·0	86·0	..	68·00	9·77	12·67	22·44

Banana.—(*Concluded.*)

Name. मराठी	Skin.	Edible Matter.	On Edible matter %	Moisture.	Reducing.	Non-reducing.	Total Sugars.
Velehi	13·0	87·0	..	65·00	10·42	10·24	20·66
Mutheli	15·82	84·18	..	76·00	15·50?	4·60?	20·10
Rajeli	13·99	86·01	..	76·00	10·20	7·14	17·34
R ·swal	23·00	77·00	..	68·00	4·50	14·00	18·50

Detailed analysis of edible portion of some varieties of Bananas.

	Phosphoric acid (P_2O_5).	Nitr.	Moisture.	Reducing.	Non-reducing.	Total Sug.
Sonkeli	0·14	0·21	66	17·5	3·2	20·7
Velchi	0·15	0·19	62	22·63	5·10	27·73
Rojeli	0·18	0·19	76·6	15·20	4·30	19·50

Proximate Composition of some Hawaiian Fruits, similar to Indian fruits.

Name.	Water %.	Protein (N × 6·25).	Ether extract (Fat.)	Crude Fibre.	Carbo-hydrates (by difference.)	Ash.	Acid (as citric.)
Banana	65—71	0·9—1·5	0·2—0·3	0·4—0·7	26—32	0·76—0·8	0·36—0·6
Coco-nut—Water only and no meat.	95·8	..	..	..	..	0·5	
Coco-nut—Water from nuts with soft spoon meat.	93·8	..	..	..	..	0·4	
Litchi	77—86	0·8—0·9	0·2—0·3	0·2	11·8—20·6	0·3—0·37	..
Mango, Pirie	80	0·5	0·2	0·7	18·2	0·37	0·97
Papaya	85·6	0·5	0·3	0·8	12·3	0·51	0·13
Pineapple	86·0	0·5	0·2	0·5	13·0	0·28	0·48
Strawberry	90·5	0·8	0·2	1·0	7·1	0·43	..
Watermelon	90·2	0·9	0·1	0·1	8·3	0·37	0·09
Tamarind	33·9	3·3	0·5	1·8	*45·7	2·57	12·16

* Calculated as percent by difference including acids as citric.

Nutritive Value of Foods.

Food.	"Good" protein.	Minerals.	Vitamins A	B	C	D
Milk	* *	* * *	*	*	*	*
Cheese	* *	* *	*	*	—	—
Eggs	* *	* *	*	**	—	**
Liver	* *	* *	*	**	—	*
Fat fish	*	..	*	*	—	**
Green vegetables, salads ..	*	* * *	*	*	**	—
Raw fruit, fruit juices ..	..	* * *	*y	*	**	—
Butter	—	—	*	—	—	*
Cod-liver oil	—	—	***	—	—	***
Meat (muscle)	*	t	—	*	t	—
Root vegetables, tubers ..	..	..	*y	*	*	—
Legumes (dry peas, lentils) ..	..	..	—	*	—	—
Cereals, bread (whole-meal) ..	*	t	t	*	—	—
Cereals, rice (polished) ..	—	—	—	—	—	—
Nuts	t	..	—	**	—	—
Sugar, jam, honey	..	..	—	—	—	—
Vegetable Oils	..	..	—	—	—	—

Symbols used; * * * stands for very rich; * * stands for rich; * signifies present; t means traces; y means if yellow in colour.

Table of Food-values and their prices.

	No. of Cal. per seer of the product.	Price in pies per seer of the product.	Price per 100 Cal. in pies.
Milk	616	48	7.7
Roasted Beef ..	1632	144	8.8
Mutton	1722	144	8.3
Chicken	2176	200	9.1
Fish	832	100	12.0
Cabbage, Turnips (boiled).	128	12	9.3
Carrots	158	15	9.5
Tomatoes	I92	48	25.0
Lentils-Peas (dried) ..	1024	60	5.8
Eggs with shell (av. lot 58 gms.)	1254	165	13.1
Potatoes	294	24	8.1
Bread	2386	60	2.5
Rice (boiled).. ..	1152	48	4.1
Oranges, Lemons (fresh)	448	75	17.1
Apples, Pears ..	640	120	18.7
Grapes	832	144	17.3

	Ozs.	Gms.	Prot.	Fat.	Carbo-hydrates.	Cal.
Dates	8	227	4.0	5.6	160.8	710
Figs 1 Fig	=1	28.4	1.3	0.1	22.2	95
	8	227	9.7	0.7	168.2	718
Horlick's Malted milk	½	14	2.3	1.2	9.5	59.5
Onion ..	4	113	1.8	0.3	11.2	56.0
Orange ..	5	142	1.2	0.3	17.4	77
Pea-nuts (Shelled). ..	5	142	36.5	54.7	34.5	777
Peas (green) ...	4	113	7.7	0.5	19.6	114
Rice ..	½	15	1.1	0.04	11.2	50
Tomato ...	½	15	0.2	0.03	0.6	4
Walnuts (English) ..	5½	156	25.8	98.8	25.1	1093

Percentage composition of
Animal Food-stuffs.

Food-stuff.	Water.	Albuminoids.	Fat.	Carbo-hydrates.	Salts.
Woman's Milk	89.2	2.1	3·4	5·0	0.2
Cow's Milk	87·5	3·4	3·6	4·8	0·7
Butter	14·15	0·68	83·11	0·7	1·2
Cheese (fatty) ..	35·75	27·15	30·45	2·5	4·15
Cheese (half-fatty) ...	46·8	27·1	20·6	2·0	3·1
Cheese (thin)	48·0	32·7	8·4	6·8	4·1
Skimmed Cow-milk ...	90·55	3·1	0·8	4·8	0·75
Ox-flesh (medium fat) ...	72·25	21·4	5·2	..	1·15
Calf-flesh (thin) ..	78·8	19·9	0·8	..	1·2
Pork (medium fat) ..	64·5	14·5	20·0	..	1
Mutton (medium fat) ..	70	17	6	..	1
Liver	72·5	19·5	6	..	2
Ham (smoked) ..	28·0	24	36·5	1·5	10
Haddock (fish) ..	80·9	17·1	0·4	..	1.6
Herring (fresh) ..	80·7	10·1	7·1	..	2·1
Herring (salted) ..	47·2	18·9	16·9	...	16.5
Eggs	74	14	11	...	1

(From Prof. H. Bischoff).

Vegetable Food-stuffs.

% Composition.	Water.	Albu-miniods.	Fat.	Carbo-hydrates.	Wood fibres.	Salt.
Wheat	13·6	12·5	1·7	67·9	2·7	1·8
Rye	15·3	11·4	1·7	67·8	2·0	1·8
Wheat flour (fine) ...	14·9	8·9	1·1	74·2	0·3	0·6
Rye-flour ..	14·2	11·0	2·0	69·8	1·6	1·5
Semolina ...	15·1	11·7	1·7	70·9	0·1	0·5
Wheat-bread (fine) ...	38·1	6·8	0·8	43·3	0·4	1·2
Ryebread (fresh) ...	42·3	6·1	0·4	49·2	0·5	1·5
Rice (husked) ..	12·6	6·7	0·9	78·5	0·5	0·8
Beans	13·5	25·3	1·7	48·3	8·1	3·1
Peas	13·9	23·2	1·9	52·7	5·7	2·7
Lentil	12·3	25·9	1·9	52·8	3·9	3·0
Potatoes ..	75·0	2·1	0·2	21·2	0·7	1·1
Carrots	86·8	1·2	0·3	9·2	1·5	1·0
Cucumbers ..	95·2	1·2	0·1	2·3	0·8	0·4
Asparagus ..	93·8	1·8	0·25	2·7	1·0	0·5
Green Peas ..	78·4	6·4	0·53	12·0	1·9	0·8
Green Beans ..	84·1	5·4	0·3	7·4	2·1	0·75
Cabbage ..	90·0	2·0	0·2	4·5	2·0	1·3
Plums (fresh) ..	84·9	0·4	..	9·8	4·3	0·6
Cocoa	5·55	20·3	28·35	39·5	..	6·5
Chocolate ..	1·6	7	22·2	64·9	..	2·3

(From Prof. H. Bischoff).

BIBLIOGRAPHY.

1. Methoden zur Untersuchung von Milch und Milch-erzeugnissen, Second Edition, by K. Teichert (1927).

2. Milch, Butter und Kaese, von Prof. Dr. Heinrich Luers (1923).

3. Food, by Robert McCarrison (1931).

4. The Chemical Composition of the Food Grains, Vegetables and Fruits of Western India, by D. L. Sahasrabuddhe, M. Ag., M. Sc., Poona, (1925).

5. Ernaehrung und Nahrungsmittel, von Prof. H. Bischoff, (1921).

6. Some Fruits of Hawaii, Bulletin No. 77 (1936).

7. Milchwirtschaftliches Taschenbuch, Berlin (1936).

8. Soya Bean, by F. S. Kale, Baroda, (1936).

9. Die Milch das beste und billigste Nahrungsmittel von Karl Junger (1927).

10. Butter-fat its composition, nutritive value, etc., by N. N. Godbole and Sad Gopal, B. H. University, (1930).

11. A comparative study of the utility of the Milks and Ghees of the Indian Cow and Buffalo as human food Materials, with a preface by Mahatma Gandhi, Jnanmandal Press, Benares, (1935).

12. Kohlenwasserstoffoele und Fette, Seventh Edition, von Holde-Bleyberg (1933).

13. Jahrbuch der Milchwirtschaft, von Dr. K. Teichert, (1934).

14. Practical Dietics with reference to Diet in Disease, by Alida Frances Pattee, Mount Vernon, New York, (1914).

15. Milk Supply, in Lahore, (1930)

16. Industrial Chemistry, Organic, by Martin, (1913).

17. The Problem of Nutrition, Vol. I., published by the League of Nations, June (1936).

18. Kautilya's Arthashastra, original and Translation by R. Shamsastry, (1923).

19. Live-stock Statistics, published by the Government of India, (1929-1930) and (1935).

20. Rig-Veda and Atharva Veda and Ayurvedic References.

21. News-papers, Monthlies, Weeklies etc.

INDEX OF SUBJECT AND NAMES.

A

B

C

D

E

R

S

T

U

V

W

X

Y

Z

AUTHOR INDEX.

BOTANICAL NAMES.